Oxford KS3 Science

Activate
Question • Progress • Succeed
3

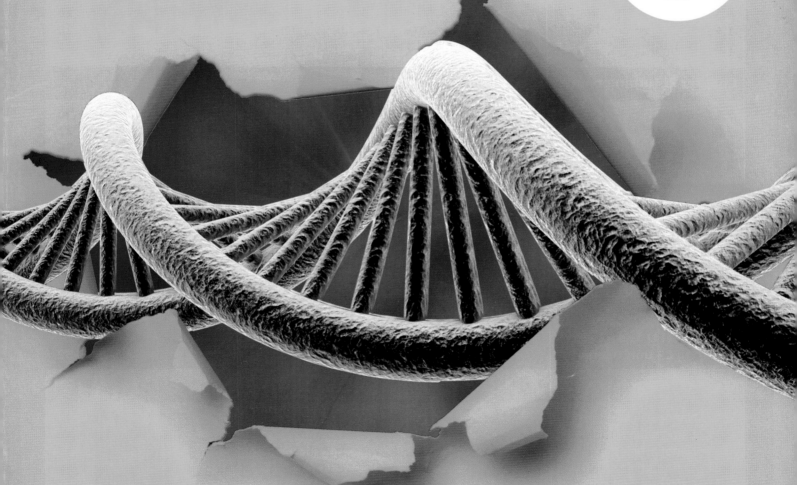

Philippa Gardom Hulme
Jo Locke
Helen Reynolds

Assessment Editor
Dr Andrew Chandler-Grevatt

OXFORD
UNIVERSITY PRESS

Contents

| Introduction | IV |

Biology B3
Biology B3 Unit Opener — 2

Chapter 1: New technology

1.1	Genetics	4	1.6	Biotechnology 1	14
1.2	Inherited disorders	6	1.7	Biotechnology 2	16
1.3	Selective breeding	8	1.8	Enzymes in industry	18
1.4	Genetic engineering	10	1.9	B3 Chapter 1 Summary	20
1.5	Cloning	12			

Chapter 2: Turning points in biology

2.1	Vaccines 1	22	2.5	DNA	30
2.2	Vaccines 2	24	2.6	Charles Darwin	32
2.3	Antibiotics 1	26	2.7	Preventing extinction	34
2.4	Antibiotics 2	28	2.8	B3 Chapter 2 Summary	36

Chapter 3: Detection

3.1	Microscopy	38	3.5	Time of death	46
3.2	Fingerprinting	40	3.6	Pathology	48
3.3	DNA fingerprinting	42	3.7	B3 Chapter 3 Summary	50
3.4	Blood typing	44			

Chemistry C3
Chemistry C3 Unit Opener — 52

Chapter 1: New technology

1.1	Nanoparticles	54	1.6	New fuels	64
1.2	Using nanoparticles	56	1.7	Cleaning up exhausts	66
1.3	Nanoparticles in medicine	58	1.8	Hybrid electric cars	68
1.4	Nanoparticle safety	60	1.9	C3 Chapter 1 Summary	70
1.5	Cars: pros and cons	62			

Chapter 2: Turning points in chemistry

2.1	Evidence for atoms	72	2.4	Lessons from fossils	78
2.2	Looking into atoms	74	2.5	The oldest primate	80
2.3	Discovering the Periodic Table	76	2.6	C3 Chapter 2 Summary	82

Chapter 3: Detection

3.1	Break-in!	84	3.5	Body!	92
3.2	Smelly problem	86	3.6	Clues in the carpet	94
3.3	Message in a bottle	88	3.7	A week in court	96
3.4	Blood alcohol	90	3.8	C3 Chapter 3 Summary	98

Physics P3
Physics P3 Unit Opener — 100

Chapter 1: New technology

1.1	Your phone	102	1.5	Your sports	110
1.2	Your house	104	1.6	Your planet	112
1.3	Your hospital – intensive care	106	1.7	P3 Chapter 1 Summary	114
1.4	Your hospital – seeing inside	108			

Chapter 2: Turning points in physics

2.1	Discovering the Universe 1	116	2.6	Radioactivity 1	126
2.2	Discovering the Universe 2	118	2.7	Radioactivity 2	128
2.3	The Big Bang	120	2.8	Electromagnetism 1	130
2.4	Spacecraft and satellites	122	2.9	Electromagnetism 2	132
2.5	Mission to the Moon	124	2.10	P3 Chapter 2 Summary	134

Chapter 3: Detection

3.1	Detecting planets	136	3.4	Detecting messages	142
3.2	Detecting alien life	138	3.5	Detecting particles	144
3.3	Detecting position	140	3.6	P3 Chapter 3 Summary	146

Glossary	148
Index	154
Periodic Table	157

Introduction

Learning objectives
Each spread has a set of learning objectives. These tell you what you will be able to do by the end of the lesson.

Key Words
The key words in each spread are highlighted in bold and summarised in the key-word box. They can also be found in the Glossary.

Link
Links show you where you can learn more about something mentioned in the topic.

Summary Questions
1. 🧪 Questions with one conical-flask symbol are the easiest.
2. 🧪🧪 The questions get harder as you move down the list.
3. 🧪🧪🧪 The question with three conical-flask symbols is the hardest. In these questions you need to think about how to present your answer.
In QWC questions you need to pay attention to the Quality of Written Communication.

Welcome to your *Activate* Student Book. This introduction shows you all the different features *Activate* has to support you on your journey through Key Stage 3 Science.

As you work through this Student Book, you'll meet and build on some familiar concepts, as well as discovering new scientific ideas.

This book is packed full of fantastic (and foul!) facts, as well as plenty of activities to help build your confidence and skills in science.

Q These boxes contain short questions. They will help you check that you have understood the text.

Maths skills
Scientists use maths to help them solve problems and carry out their investigations. These boxes contain activities to help you practise the maths you need for science. They also contain useful hints and tips.

Literacy skills
Scientists need to be able to communicate their ideas clearly. These boxes contain activities and hints to help you build your reading, writing, listening, and speaking skills.

Working scientifically
Scientists work in a particular way to carry out fair and scientific investigations. These boxes contain activities and hints to help you build these skills and understand the process so that you can work scientifically.

Fantastic Fact!
These interesting facts relate to something in the topic.

Opener
Each unit begins with an opener spread. This introduces you to some of the key topics that you will cover in the unit.

You already know
This lists things you've already learnt that will come up again in the unit. Check through them to see if there is anything you need to recap on.

Big questions
These are some of the important questions in science that the unit will help you to answer.

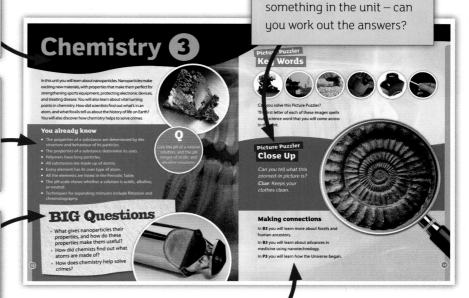

Picture Puzzlers
These puzzles relate to something in the unit – can you work out the answers?

Making connections
This shows how what you will learn in the unit links up with the science that you will learn in other parts of the course..

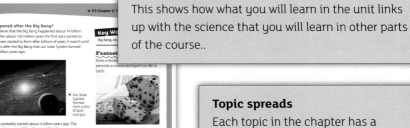

Topic spreads
Each topic in the chapter has a double-page spread containing learning objectives, practice questions, key words, and task boxes to help you work through the chapter.

Summary
This is a summary of the chapter. You can use it to check that you have understood the main ideas in the chapter and as a starting point for revision.

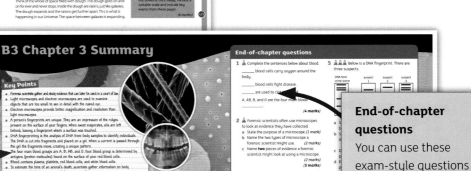

End-of-chapter questions
You can use these exam-style questions to test how well you know the topics in the chapter.

Big write/Maths challenge/Case study
This is an activity that you can do at the end of the chapter. It will help you to practise using your scientific skills and knowledge.

Biology 3

In this unit, you will begin by looking at genetics. This includes genetically inherited disorders and how plant and animal genes can be changed to alter an organism's characteristics. You will also study how you can protect yourself from disease through immunisation and treat conditions using antibiotics. Finally, you will find out how forensic scientists help to solve crimes through the analysis of evidence found at the scene of a crime.

You already know

- Plants and animals, including humans, resemble their parents and share many features.
- Genetic information is passed from one generation to the next.
- Living organisms produce offspring of the same kind but normally offspring vary and are not identical to their parents. They exhibit variation.
- Animals and plants are suited to and adapt to their environment in different ways.
- Variation and adaptation lead to evolution.
- Living organisms have changed over time and fossils provide evidence and information about living things that inhabited the Earth millions of years ago.

Q What is meant by the term habitat?

BIG Questions

- How can we create new food products?
- How can we protect ourselves against disease?
- How do forensic scientists help to solve crimes?

Picture Puzzler
Key Words

   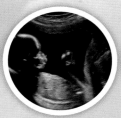 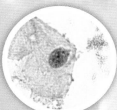 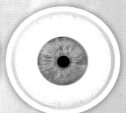

Can you solve this Picture Puzzler?

The first letter of each of these images spells out a science word that you will come across in this unit.

Picture Puzzler
Close Up

Can you tell what this zoomed-in picture is?

Clue: This natural plant fibre was found at a crime scene.

Making connections

In **B3** you will learn about advances in medicine.

In **C3** you will learn more about fossils and human ancestors.

In **P3** you will learn how the Universe began.

1.1 Genetics

Learning objectives

After this topic you will be able to:
- describe the difference between dominant and recessive alleles
- use a Punnett square to show what happens during a genetic cross.

Have you ever wondered why brothers and sisters often have a similar appearance? It's all down to the genes they inherit from their parents.

Which characteristics will you inherit?

For each characteristic, you have two genes. One gene is inherited from your mother, and one from your father. These two genes may be the same, or different. Different forms of the same gene are called **alleles**.

A State what is meant by an allele.

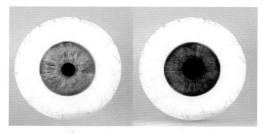

▲ The gene for eye colour has an allele for blue eye colour and an allele for brown eye colour.

How is eye colour inherited?

Some alleles will always produce a characteristic in an organism. These are called **dominant** alleles. You only need one copy of a dominant allele for the characteristic to appear in the organism. This allele is said to be 'expressed' in the organism.

B Name the type of allele that will always produce a characteristic in an organism.

Link

You can learn more about genes in B2 3.5 Inheritance

Key Words

allele, dominant, recessive, Punnett square

For example, the allele for brown eyes is a dominant allele. If you inherit this allele from your mother, your father, or both parents, you will have brown eyes.

The allele for blue eye colour is a **recessive** allele. You need two copies of a recessive allele for the characteristic to be expressed in the organism.

For example, you will only have blue eyes if you inherit this allele from both your mother and your father.

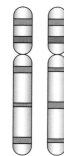

▲ To have blue eyes, you need two copies of the allele for blue eyes.

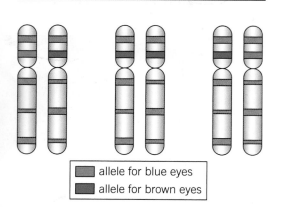

allele for blue eyes
allele for brown eyes

▲ These genes belong to people who all have brown eyes.

C State how many copies of a recessive allele are needed for it to be expressed.

B3 Chapter 1: New technology

Can characteristics be predicted?
When a sperm fertilises an egg, genes from the mother join with genes from the father. This results in the combination of alleles present in the offspring. Scientists are able to predict what an organism's offspring will look like by carrying out a genetic cross.

How do you perform a genetic cross?
In a genetic cross, alleles are represented by letters. The dominant allele is represented by a capital letter, and the recessive allele by the same, lowercase letter.

When studying eye colour, 'B' is used to represent the dominant allele for brown eyes, and 'b' represents the recessive allele for blue eyes.

Scientists use a **Punnett square** to show what happens to the alleles in the genetic cross. In this example, a mother with blue eyes (bb) is crossed with a father with brown eyes (BB).

Mother: blue eyes

Father: brown eyes
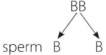

A sperm and egg cell only contain one copy of each gene.

A Punnett square is actually a simple table. To produce a Punnett square, put the possible alleles from one parent across the top of the square, and the alleles from the other parent down the side.

		Father	
		B	B
Mother	b		
	b		

Use the square to work out the possible combinations of alleles in the offspring.

In this example, all offspring produced will have brown eyes. This is because the dominant allele is present in all possible combination of the parents' alleles.

		Father	
		B	B
Mother	b	Bb	Bb
	b	Bb	Bb

If the father's alleles are Bb, he will still have brown eyes but now it is possible that the offspring will have blue eyes.

		Father	
		B	b
Mother	b	Bb	bb
	b	Bb	bb

In this example, two of the four combinations are bb, which means there is a 2 in 4 chance that the offspring will have blue eyes, and a 2 in 4 chance it will have brown eyes.

Genetic-cross outcomes
Scientists often display the possible outcomes from a genetic cross as the probability of a characteristic being expressed. This could be in the form of a ratio, a percentage, or a fraction. For example, a 1 in 5 probability is 1/5 = 0.2 = 20%.

Write each of the following as a fraction and a percentage:

0 in 4   1 in 4   2 in 4
3 in 4   4 in 4

Summary Questions

1. Copy and complete the sentences below.
   Different forms of the same gene are known as _____ . _____ will always be expressed if they are present. _____ alleles will only be expressed if two copies are present.
   (3 marks)

2. In mice, black fur is dominant and white fur is recessive. State the fur colour a mouse would have with the following allele combinations:
   a BB   b bb   c Bb
   (3 marks)

3. Use a Punnett square to calculate and explain the chance of a person inheriting freckles if their mother has the alleles Ff and their father has the alleles Ff. Freckles are a dominant characteristic.
   (6 marks)

1.2 Inherited disorders

Learning objectives

After this topic you will be able to:
- describe what is meant by a genetically inherited disorder
- calculate the probability of a person suffering from an inherited disease.

▲ Genetic counsellors work out the chance of a couple's child being born with an inherited disorder.

You can become ill by catching diseases such as coughs and cold from other people. Other medical conditions can be inherited in a person's genes.

What is a genetically inherited disorder?

Genetically inherited disorders are conditions passed from parents to their offspring in their genes. Examples include cystic fibrosis, haemophilia, and polydactyly.

A Write down a definition of the term genetically inherited disorder.

What is cystic fibrosis?

Cystic fibrosis sufferers produce lots of thick sticky mucus. This blocks their air passages, making it difficult to breathe, and can lead to chest infections. The excess mucus also causes difficulty in food being absorbed. Physiotherapy and antibiotics help to manage the symptoms but there is no cure.

Cystic fibrosis is caused by a recessive allele, so you need two copies of the allele to have the disorder. If either of your genes contains a copy of the healthy dominant allele, you will not suffer from cystic fibrosis.

If a person has one copy of the dominant allele and one copy of the recessive allele, they are called a **carrier**. This means that they carry a copy of the allele, but do not have the disorder.

B State what is meant by a carrier.

What is the chance of inheriting cystic fibrosis?

Genetic counsellors can use a Punnett square to determine the chance of a child inheriting a condition from their parents. If the chance is high, couples may decide against having a child.

In this example, c is used to represent the allele for cystic fibrosis. The healthy (dominant) allele is represented with the letter C.

Mother: carrier

eggs

Father: carrier
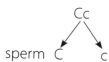
sperm

> **Genetically inherited disorders**
> Produce a presentation to show the main symptoms, possible treatments, and likelihood of suffering from cystic fibrosis.

The Punnett square shows the possible combinations of alleles in the offspring:

	Father	
	C	c
Mother C	CC	Cc
Mother c	Cc	cc

The outcomes are:
- CC – one healthy
- Cc – two carriers
- cc – one cystic fibrosis sufferer

Out of the four possible outcomes, there is a 1 in 4, or 25%, chance of a child from this couple suffering from cystic fibrosis.

What is polydactyly?

Polydactyly is a disorder that results in a child being born with extra digits on their hands or feet. These can be small stumps of soft tissue, or extra fingers or toes.

Polydactyly is caused by a dominant allele. Therefore, if one of your genes contains this allele, you will have the disorder. You cannot be a carrier of this disorder.

c State why it is not possible to be a carrier of polydactyly.

What is the chance of inheriting polydactyly?

The genetic cross below shows the likelihood of a person suffering from polydactyly if the mother has the disorder but the father does not. In this example, P is used to represent the allele for polydactyly.

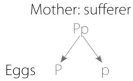
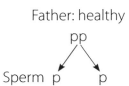

The Punnett square shows the possible combinations of alleles in the offspring:

	Father	
	p	p
Mother P	Pp	Pp
Mother p	pp	pp

The outcomes are:

pp – two healthy

Pp – two polydactyly sufferers

Out of the four possible outcomes, there is a 2 in 4, or 50%, chance of a child from this couple suffering from polydactyly.

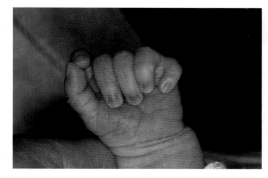

▲ This suffer of polydactyly has an extra finger.

Key Words

genetically inherited disorder, carrier

Summary Questions

1. Copy and complete the sentences below.

   Genetically _____ disorders are passed on from _____ to their _____ through their genes. The disorders can be caused by dominant or recessive alleles. If a disorder is _____ you need two copies of the allele to suffer from the disorder. If a person has only one copy of the allele, they are called a _____.

   (5 marks)

2. Explain why most people would not know that they are a carrier of a genetically inherited disorder.

   (3 marks)

3. Use a Punnett square to explain why a child could not suffer from cystic fibrosis if his mother's alleles were CC and his father's were Cc.

   (6 marks)

1.3 Selective breeding

Learning objectives
After this topic you will be able to:
- describe the process of selective breeding
- describe some advantages and disadvantages of selective breeding.

Link
You can learn more about breeding in B2 3.6 Natural selection

Key Words
selective breeding

Milk production
Produce an information leaflet to show farmers the steps involved in selectively breeding cows for milk production.

You can often guess what product a farmer is rearing their livestock for. Large herds of dairy cows are used for milk production whereas sheep with long, thick coats are reared for wool. Farmers can increase their production levels by using a technique known as selective breeding.

A State what is meant by the term selective breeding.

What is selective breeding?
Most farmers choose the animals or plants they raise for their characteristics. For example, a farmer may select dairy cattle that produce lots of milk, or strawberry plants that produce lots of large berries.

When producing offspring, the farmer will choose their best plants or animals to breed. This is **selective breeding**. The offspring produced are likely to share their parents' desirable characteristics.

B State an advantage of selective breeding.

How does a farmer selectively breed organisms?
There are five main steps in selectively breeding an organism. These are the same for both plants and animals.

1. Decide which characteristic(s) of the species is most important.
2. Select parents that show high levels of this characteristic.
3. Breed these individuals.
4. Select the best offspring and breed again.
5. Repeat for many generations.

Disadvantages of selective breeding
When you selectively breed an organism, you are choosing which versions of a gene are passed on. By making organisms look more and more similar over each generation, you are reducing the number of genes (the gene pool) from which a species is created. Selective breeding therefore reduces variation within a species.

C State the effect of selective breeding on variation.

● B3 Chapter 1: New technology

Selectively breeding sheep

Step 1
Decide which characteristic is most important. In this example, the farmer wishes to select sheep that produce large, good-quality fleeces.

Step 2
Select parents that show high levels of this characteristic.

Step 3
Breed these individuals.

◀ Selectively breeding sheep for wool production.

Step 4
Select the best offspring. Breed again from this generation.

Step 5
Repeat the process over many generations. Eventually, all sheep will have large, good-quality fleeces.

'Useful' genes, which may be needed in the future, could be lost. For example, if a new disease occurs, an organism may not exist that contains the gene for resistance to this disease. This could result in a species becoming extinct.

D State a disadvantage of selective breeding.

Selectively breeding dogs

Pedigree dogs are selectively bred so that they display the best characteristics of their breed. This could be for speed and strength, or appearance. However, many suffer from health problems as a result of this breeding. They are at higher risk of genetically inherited disorders.

▲ Pedigree dogs.

▲ Pugs are selectively bred to have short, stubby noses. This can cause breathing problems.

For example, many Labradors suffer from hip problems and pugs suffer from breathing problems.

The chance of inheriting a genetic defect is increased when animals are in-bred, for example, when closely related dogs, such as brother and sister, are bred. One result of in-breeding is that pedigree dogs have a much lower life expectancy than crossbreeds.

Summary Questions

1. Re-arrange the steps below into the correct order to show how organisms are selectively bred.

   Repeat the process for many generations.

   Select parents that show high levels of the desirable characteristic.

   Select the best offspring and breed again.

   Breed these individuals.

   Choose a desirable characteristic.

   (5 marks)

2. Describe an advantage and disadvantage of selectively breeding dogs.

   (2 marks)

3. A farmer wants to produce large sweet tomatoes. He currently grows two species of tomato:
   - One species produces very sweet tomatoes.
   - The other species produces large tomatoes but they are not very sweet.

   Explain in detail how the farmer can selectively breed the two species to produce large sweet tomatoes.

   (6 marks QWC)

1.4 Genetic engineering

Learning objectives

After this topic you will be able to:
- state how a product is produced using genetic engineering
- describe some advantages of producing products through genetic engineering.

▲ GloFish.

Have you ever seen fish that glow in the dark? Scientists have altered the genes of one type of fish to make them fluoresce (glow) by genetic engineering. The aim was to produce a fish that would glow in the presence of polluted water.

What is genetic engineering?

When farmers selectively breed plants and animals, they are choosing organisms' genes. However, this is a slow process that takes place over many generations. It is also not very precise.

Scientists are now able to alter an organism's genes to produce an organism with desired characteristics. For example, crops can be produced that are resistant to disease. This is called **genetic engineering** (or genetic modification).

A State what is meant by genetic engineering.

This is a very precise process, as single genes can be targeted. It can also happen in one generation so is a much quicker process than selective breeding.

B State an advantage of genetic engineering.

Examples of genetic engineering

Many organisms have been genetically engineered. For example:
- cotton – to produce high yields
- corn – to produce toxins (poison) that kill insects
- bacteria – to produce medicinal drugs.

How can you alter an organism's genes?

To create an organism with a desired characteristic, scientists take genes from another organism that shows this characteristic. These are known as foreign genes. The foreign genes are put into plant or animal cells at a very early stage of the organism's development. As the organism develops, it will display the characteristics of the foreign genes.

Genetic engineering cartoon strip

Select one example of genetic engineering. Produce a cartoon strip that explains simply how an organism can be genetically engineered to produce a desired characteristic. Write a short caption for each step of the cartoon strip.

Frost-resistant tomatoes

The flounder is a fish that lives in very cold waters, and contains a gene to prevent it freezing. Scientists have created frost-resistant tomatoes by inserting the flounder's antifreeze gene into the cells of a tomato plant. This type of genetically engineered tomato plant is no longer destroyed by frost, which is very beneficial for farmers.

Key Words
genetic engineering

▲ Flounders produce antifreeze chemicals that allow them to live in very cold water.

▲ Tomatoes can be made frost resistant by adding the flounder's antifreeze gene.

Genetically engineered bacteria

Bacteria can be genetically engineered to produce many useful chemicals, including vaccines and antibiotics. As bacteria reproduce very quickly, they can be used to produce large amounts of the chemical in a very short period.

C Name two useful chemicals produced by genetically engineered bacteria.

Your body needs a chemical called insulin to control your blood–sugar level. Some people do not produce enough insulin, and must inject it daily. The insulin they need can be made using genetically engineered bacteria:

1. Genes that code for the production of insulin are inserted into the bacteria.
2. The bacteria now produce insulin.
3. The bacteria multiply many times, and produce large quantities of insulin.
4. The bacteria are then removed, leaving behind the useful insulin.

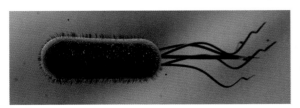

◀ *E.coli* is genetically engineered to produce insulin.

Summary Questions

1. Copy and complete the sentences below.

   Scientists can insert _____ genes into organisms to produce desired _____. This is called _____ engineering. For example, bacteria can be engineered to produce _____.

   (4 marks)

2. Describe the advantages of genetic engineering over selective breeding.

   (3 marks)

3. Haemophilia is a disease that prevents blood from clotting. Sufferers of this disease are unable to produce Factor VIII, a chemical that clots blood. Describe how bacteria can be genetically engineered to produce large quantities of Factor VIII for the treatment of haemophilia.

   (6 marks QWC)

1.5 Cloning

Learning objectives

After this topic you will be able to:
- describe what is meant by a clone
- describe some advantages and disadvantages of cloning.

Link

You can learn more about the way plants reproduce in B1 3.6 Flowers and pollination

Can you tell the difference between two blackbirds? Perhaps you could if you looked carefully. Can you tell the difference between two bacteria? Probably not. This is because bacteria produce identical copies of themselves.

What is a clone?

A **clone** is an organism that is genetically identical to its parent. This means that it has the same genes as its parent, and will therefore look identical.

When bacteria reproduce, they divide in half. Each half contains identical genes, resulting in two identical organisms, called clones. The production of clones is useful when bacteria are used to make chemicals such as insulin.

A State what is meant by a clone.

Asexual reproduction

Bacteria do not reproduce using a partner. Only one parent is needed; this is known as **asexual reproduction**. New organisms are created by cell division. There is no mixing of genetic material so the offspring produced are clones.

Many plants can also reproduce asexually. For example:

- potato plants – produce many tubers, each of which can grow into a new plant
- strawberry plants – produce long stems with tiny plants (plantlets) on the end
- daffodils – at the end of each growing season a new bulb is formed from which next year's flower grows.

▲ Spider plants reproduce asexually by producing plantlets on side branches.

B State what is meant by asexual reproduction.

How can you make clones?

When you take a plant cutting, you are making a clone. Most plants can be grown from cuttings. A cutting is a small section of a plant. It is planted and allowed to grow into a new plant. Often the

Key Words

clone, asexual reproduction

B3 Chapter 1: New technology

cutting is dipped in a special chemical called rooting powder to encourage the cutting to grow roots.

Advantages of this technique are:
- new plants are produced quickly
- the technique is cheap
- all plants are genetically identical so they will all have the desired characteristics.

The main disadvantage is that cloning plants reduces the variety of genes available (the gene pool). This can increase the risk of disease, or a change in the plant's environment, destroying a species.

C State a disadvantage of cloning plants.

How can you make animal clones?

Most people are happy about cloning plants but there are very mixed views on cloning animals. Some people think that cloning animals is unethical.

Two animal cloning techniques that scientists regularly use are:
- cell cloning – scientists clone human cells in the laboratory, and use them for research into diseases
- tissue culture – new skin and cartilage can be grown in a sterile environment; this technique is used to grow new skin for burns victims.

Scientists believe that there may be many beneficial uses of cloning. For example, endangered species could be cloned, preventing their extinction.

▲ Rooting powder encourages roots to grow.

Plant cuttings

A new plant can be grown by taking a cutting from another plant. This is an example of cloning.

Try taking your own cuttings to see if you can clone a plant.

▲ Dolly was the first mammal to be cloned using cells from an adult sheep.

▲ This is not actually an ear, it is just cartilage growing around a mould. Scientists hope they will be able to regrow noses and ears in the future.

Summary Questions

1 Copy and complete the sentences below.

Organisms that are genetically _____ to their parents are known as _____ . This is an example of _____ reproduction.

(3 marks)

2 Describe the main differences between asexual and sexual reproduction.

(3 marks)

3 Using examples to illustrate your answer, compare the main advantages and disadvantages of cloning.

(6 marks QWC)

1.6 Biotechnology 1

Learning objectives

After this topic you will be able to:
- write the word equation for fermentation
- describe how bread, beer, and wine are made.

Many of the food and drinks we consume have been made using microorganisms. For example, yeast is added to bread to make it rise. This is an example of biotechnology.

What is biotechnology?

Biotechnology is the use of biological processes or organisms to create useful products. Many of these products are foods and drinks.

A State what is meant by the term biotechnology.

What is yeast?

Yeast is a microorganism. It is used in the production of bread and many alcoholic drinks. These products are made using the chemical reaction **fermentation**. Fermentation is a type of anaerobic respiration – the yeast respires without needing oxygen.

Fermentation can be represented by the following word equation:

glucose → ethanol + carbon dioxide (+ energy)

B Write down the word equation for fermentation.

Enzymes present in the yeast speed up fermentation, making the reaction occur faster. The enzymes work best in a warm environment.

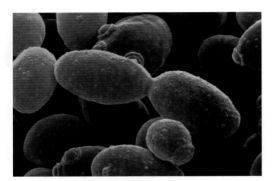

▲ *Saccharomyces cerevisiae* is the yeast used to make bread.

How do you make bread?

Flour, water, and yeast are mixed to make dough. The dough is then left in a warm place to rise. This is caused by the yeast respiring, changing the sugars in the flour into ethanol and carbon dioxide. The carbon dioxide gas is trapped as bubbles inside the dough, making it rise.

The dough is then baked. In the oven, the ethanol evaporates. The bubbles of gas expand, making the bread rise further.

▲ Before baking, the bread is left to rise.

C Name the gas that makes dough rise.

Link

You can learn more about fermentation in B2 2.6 Anaerobic respiration

● B3 Chapter 1: New technology

How do you make beer and wine?

▲ Alcoholic drinks are made by fermenting plant sugars.

Beer and wine are made in very similar ways. The type of alcoholic drink produced depends on the source of sugar. This determines the type of ethanol produced.

Wine is made when yeast is used to ferment grape sugar. Beer or lager is made when yeast is used to ferment sugar in malted barley.

Fermenting sugar

When yeast ferments sugar, carbon dioxide is produced. Design an investigation to determine the ideal temperature for yeast to ferment sugar.

Key Words

biotechnology, fermentation

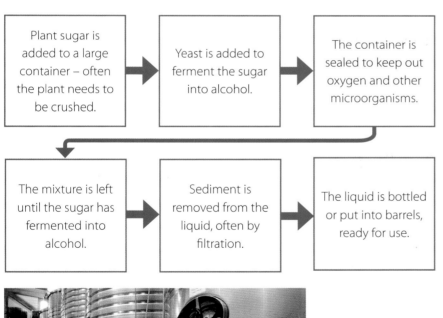

- Plant sugar is added to a large container – often the plant needs to be crushed.
- Yeast is added to ferment the sugar into alcohol.
- The container is sealed to keep out oxygen and other microorganisms.
- The mixture is left until the sugar has fermented into alcohol.
- Sediment is removed from the liquid, often by filtration.
- The liquid is bottled or put into barrels, ready for use.

◄ Wine fermenters are kept warm to speed up the process of fermentation.

Summary Questions

1. 🧪 Copy and complete the sentences below.

   Yeast is a _____. It is used to make bread and _____ drinks.

   During _____, the _____ in yeast convert glucose into ethanol and _____ _____.

   (5 marks)

2. 🧪🧪 Some types of bread are made without using yeast. Suggest and explain how these breads would differ in appearance from bread made with yeast.

   (3 marks)

3. 🧪🧪🧪 Cider is an alcoholic drink made from apples. Explain how cider could be produced.

   (6 marks QWC)

15

1.7 Biotechnology 2

Learning objectives

After this topic you will be able to:
- describe the role of bacteria in fermentation
- describe how cheese and yoghurt are made.

There are many different types of cheese. You may know that cheese is made from milk but did you know that bacteria are also used in cheese-making? Some types of cheese also need mould to be produced.

How do you make cheese?

Cheese is made from the milk of animals, including cows, goats, and sheep. Different types and flavours of cheese can be made using different species of bacteria and moulds.

To make cheese, bacteria are added to milk. The bacteria then ferment lactose, a type of sugar found in milk. During fermentation, the lactose is converted into lactic acid. This acid gives cheese its tangy taste.

▲ Stilton cheese has mould (a type of fungus) growing throughout it.

A Name the product made when lactose is fermented.

The flow chart below shows how cheese is made:

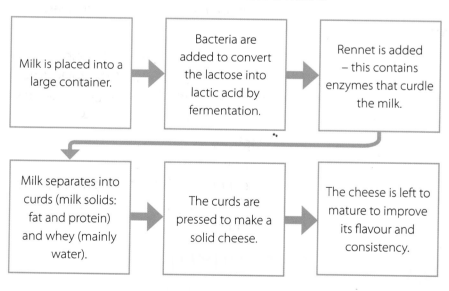

▲ You can see the curds (milk solids) in the man's hand. The watery liquid is known as whey.

B State the purpose of rennet in cheese-making.

How is yoghurt made?

Yoghurt is also made using bacteria. Normally cow's milk is used but any milk can be turned into yoghurt.

▲ Yoghurt production.

The flow chart on the next page shows how yoghurt is made:

● B3 Chapter 1: New technology

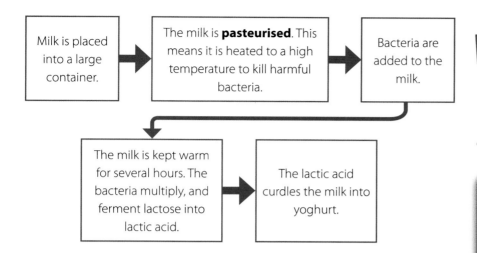

The lactic acid produced by the bacteria has another useful property. It prevents the growth of harmful bacteria. This increases the time that yoghurt can be kept and eaten safely.

C State what is meant by milk that is pasteurised.

What is 'live' yoghurt?

After production, some yoghurts are pasteurised to kill the bacteria used to make them. Yoghurts that have not been pasteurised are known as 'live' or 'probiotic' yoghurts – they contain actively growing bacterial cultures.

Pharmacists often recommend that people taking antibiotics (medicines that kill bacteria) eat 'live' yoghurt. This is to replace the useful bacteria that are present in your intestine. These are essential for digestion but are killed by any antibiotics you may take.

Some people eat probiotic yoghurts to help digestion. Normal yoghurt usually contains two types of bacteria, *Lactobacillus bulgaricus* and *Streptococcus thermophiles*. Probiotic yogurt contains additional types of bacteria that are thought to help a person's digestion, making you feel healthier.

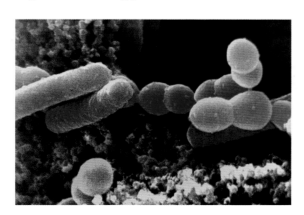

◀ *Streptococcus thermophilus* (red) and *Lactobacillus bulgaricus* (blue) are the most common types of bacteria used in yoghurt production.

Link

You can learn more about probiotic foods in B2 1.5 Bacteria and enzymes in digestion

Fermentation products

Produce an infographic explaining clearly how one food or drink product is made using the process of fermentation.

Key Words

pasteurised

Summary Questions

1 Copy and complete the sentences below.

_____ are used to _____ the sugars in milk in the production of cheese and _____. The bacteria convert _____ into lactic acid.

(4 marks)

2 Describe two useful effects of lactic acid in yoghurt production.

(2 marks)

3 Compare the processes of cheese and yoghurt production.

(6 marks QWC)

1.8 Enzymes in industry

Learning objectives

After this topic you will be able to:
- describe some commercial uses of enzymes
- describe what happens when an enzyme is denatured.

Have you ever looked closely at a washing-powder label? Many say they are biological powders. This means they contain enzymes that help to clean your clothes.

Why use enzymes?

Enzymes are a type of catalyst – they speed up reactions without being used up. Their use in industry often saves energy, for example, by allowing reactions to be carried out at lower temperatures.

Enzymes catalyse lots of reactions so they have lots of uses. These include:

- making baby food
- extracting juice from fruit
- removing stains from clothes.

Making baby food

It is hard for newborn babies to digest high-protein foods. Many baby foods are treated with proteases to break down protein into amino acids. The newborns can then absorb the amino acids, and they get the nutrients they need out of their food.

▲ Proteases are added to baby food to break down proteins, making it easier for babies to absorb the nutrients.

A State the purpose of protease in baby food.

Making fruit juice

Fruits such as apples and oranges contain pectin. Pectin makes it harder to break down the cell walls when you squeeze the fruit to release the juice. Pectinase is added to digest pectin. This makes it much easier to squeeze the fruit, releasing more juice.

Removing stains

The enzymes in biological washing powders help remove stains by breaking down the stain into water-soluble substances. These then dissolve and are washed away.

▲ Biological washing powders are used to remove stains.

Enzyme	Function	Example of stains removed
proteases	digest protein, removing protein stains	egg, blood
lipases	digest fat, removing greasy stains	butter, oil

B State the enzyme that could be used to remove a blood stain.

● B3 Chapter 1: New technology

Temperature and enzymes

The speed at which enzymes catalyse a reaction depends on the temperature. Generally, the higher the temperature, the faster the reaction. This is true up to a certain temperature; after that the enzyme is **denatured** and can no longer catalyse a reaction.

Enzymes are proteins. They have a specific shape that matches the shape of the substance they are acting on. If they are heated too much, they lose their shape and can no longer stick to the substance. The enzyme has been denatured, which means it has been permanently changed. Different enzymes are denatured at different temperatures.

The clue is in the name

You can often work out an enzyme's function from its name. Enzymes are named according to the job they do, and end in the letters 'ase'. For example:

lipases – break down lipids (fats) into fatty acids and glycerol

proteases – break down proteins into amino acids

carbohydrases – break down carbohydrates into sugar molecules

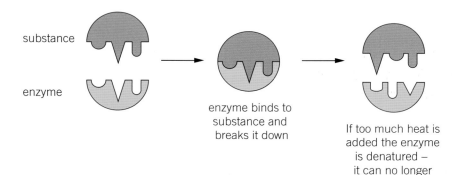

substance

enzyme

enzyme binds to substance and breaks it down

If too much heat is added the enzyme is denatured – it can no longer stick to the substance.

▲ Enzymes have to be the correct shape to bind to the substance they are acting on.

Key Words

denatured

c State what happens to an enzyme when it is denatured.

Enzyme graphs

This graph shows what happens at different temperatures to an enzyme involved in human digestion.

a Describe what happens as the temperature increases from 10 °C to 20 °C.

b What is the optimum (best) temperature for the enzyme to work at?

c Explain why the rate of digestion falls at temperatures higher than the optimum.

Summary Questions

1. Copy and complete the sentences below.

   _____ are a type of _____. They are used in reactions in industry. For example, _____ are used in washing powder to digest fat stains. If an enzyme is heated too much it will be _____.

   (4 marks)

2. Describe the use of enzymes in the production of fruit juice.

   (3 marks)

3. Explain how enzyme reactions are affected by temperature. Explain how this might affect your use of biological washing powders.

   (6 marks QWC)

19

B3 Chapter 1 Summary

Key Points

- Different forms of the same gene are known as alleles.
- Dominant alleles will always be expressed if they are present. Recessive alleles will only be expressed if two copies are present.
- Punnett squares show the possible combinations of alleles inherited from the parents. This helps scientists to predict an offspring's characteristics.
- Genetically inherited disorders are conditions passed from parents to their offspring in their genes. Examples include cystic fibrosis and polydactyly.
- A carrier has one copy of the dominant allele and one copy of the recessive allele. Carriers do not have the disorder themselves.
- Farmers selectively breed their crops and plants to produce organisms with the desired characteristics.
- Selective breeding reduces variation within a species.
- Scientists can insert foreign genes into organisms to change their characteristics. This is called genetic engineering. For example, bacteria can be engineered to produce insulin.
- Clones are organisms that are genetically identical to their parents. This is an example of asexual reproduction.
- Biotechnology is the use of biological processes or organisms to create useful products.
- Yeast is used to make bread and alcoholic drinks. During fermentation, the enzymes in yeast convert glucose into ethanol and carbon dioxide.
- Bacteria ferment milk sugars in the production of cheese and yoghurt. The bacteria convert lactose into lactic acid. This curdles the milk.
- Enzymes catalyse reactions in industrial processes. For example, enzymes are used in washing powder to digest food stains.
- If an enzyme is heated too much it will be denatured, changing its shape. It can no longer catalyse the reaction.

Key Words

allele, dominant, recessive, Punnett square, genetically inherited disorder, carrier, selective breeding, genetic engineering, clone, asexual reproduction, biotechnology, fermentation, pasteurised, denatured

Big write

Biological techniques are used in a range of different situations, including food manufacture, improving a species' characteristics, and the production of medicines.

Task
Select one example that illustrates one of the above biological techniques. Write a magazine article to explain the science used in your example, and discuss the risks and benefits.

Tips
- Use diagrams where appropriate, and explain any scientific terms you use.
- Ensure that your article is written in a 'magazine' style. For example, include images and different text styles to enhance the presentation of the work.

End-of-chapter questions

1. The following sentences describe the production of cheese. Re-arrange them into the correct order.

   Bacteria are added to convert the lactose into lactic acid by fermentation.

   The cheese is left to mature.

   Rennet is added. Enzymes curdle the milk.

   The curds are pressed to make a solid cheese.

   Milk is placed into a large container.

   Milk separates into curds.

   (6 marks)

2. Bread is made using fermentation.
   a. Circle the microorganism used in making bread.

       bacteria   yeast   virus

   (1 mark)
   b. Complete the word equation for fermentation.

   glucose → _____ + carbon dioxide (+ energy)

   (1 mark)
   c. State which of the products produced during fermentation causes bread to rise. *(1 mark)*
   d. Explain why bread does not contain ethanol. *(1 mark)*

   (4 marks)

3. A student wanted to make their own yoghurt in the laboratory.
   a. Suggest **two** safety precautions the student should take before beginning. *(2 marks)*
   b. The student's first step was to pasteurise milk. State what pasteurised means. *(1 mark)*
   c. The student then added bacteria to the milk. To encourage the bacteria to multiply, the mixture must be kept warm. Suggest how the mixture can be kept at a constant warm temperature. *(1 mark)*
   d. State the product produced when bacteria ferment milk sugars. *(1 mark)*
   e. Give two useful properties of yoghurt that contains live bacteria. *(2 marks)*

   (7 marks)

4. Enzymes are used in industry in many chemical reactions.
   a. Give **two** reasons that enzymes are used in reactions in industry. *(2 marks)*
   b. Draw and label a graph to illustrate how enzymes are affected by temperature. *(4 marks)*
   c. Describe the use of enzymes in making fruit juice. *(3 marks)*

   (9 marks)

5. Some types of tomato have been genetically engineered to stay firm for longer.
   a. Suggest why this is this an advantage for tomato sellers. *(1 mark)*
   b. Suggest **one** other characteristic that a tomato grower may choose to improve using genetic engineering. *(2 marks)*
   c. Describe how a tomato plant may be genetically engineered. *(3 marks)*
   d. Explain how genetic engineering produces plants with the desired characteristics more quickly than selective breeding. *(2 marks)*

   (8 marks)

6. Sickle cell anaemia is a genetically inherited disorder. It causes misshapen red blood cells, which can block blood vessels. The allele for this disorder is recessive. Calculate and explain the chance of a person inheriting the genetically inherited disorder if both of their parents are carriers.

   (6 marks QWC)

2.1 Vaccines 1

Learning objectives

After this topic you will be able to:
- describe the role of vaccines in fighting disease
- describe how Jenner developed the smallpox vaccine.

Have you had an injection recently? This was probably to prevent you from getting a disease like measles. These injections are known as immunisations (or vaccinations).

What is an immunisation?

To prevent you catching certain diseases you may be given an **immunisation**. This is a way of inserting a **vaccine** into your body. Most commonly this is in the form of an injection but sometimes this can be in the form of drops into the mouth.

A State what is meant by an immunisation.

What is a vaccine?

A vaccine contains dead or inactive forms of a disease-causing microorganism. This means that the microorganism can not make you ill. However, the vaccine tricks your body into thinking that the harmful, active form of the microorganism has entered your body. This triggers your **immune system** to start working. The immune system is the body's system for fighting disease. As a result of taking the vaccine, your body is protected from the disease.

B State what is meant by a vaccine.

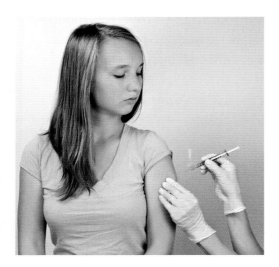

▲ This teenager is being immunised against tuberculosis. The injection contains the vaccine.

Which immunisations should you have?

Doctors encourage parents to have their children immunised at an early age. Immunisations are the most cost-effective means of preventing life-threatening infections in a population.

The table below shows the immunisations offered to every child in the UK.

Child's age	Disease immunised against
2, 3, and 4 months	polio, diphtheria, tetanus, whooping cough, Hib meningitis, meningitis C
about 13 months	measles, mumps, rubella (MMR)
3–5 years	MMR, polio, diphtheria, tetanus, whooping cough
10–14 years	tuberculosis
12-13 years	cervical cancer (girls only)
13–18 years	polio, diphtheria, tetanus

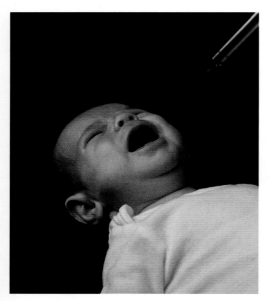

▲ The polio vaccine can be given orally.

● B3 Chapter 2: Turning points in biology

How were vaccines discovered?
The first vaccine was developed in 1796 by Edward Jenner, an English doctor. It was used to treat smallpox. Smallpox was a disease that killed one in three people who caught it, and badly disfigured those who survived.

C Name the person who developed the first vaccine.

Jenner was fascinated by news that milkmaids did not get smallpox. Most milkmaids instead suffered from a weak version of smallpox, known as cowpox. This disease caused blisters on their skin but did not cause death.

Jenner hypothesised that there must be a link between the cowpox and smallpox. He thought that the pus in the cowpox blisters somehow protected the milkmaids from smallpox. To test out his theory he took some pus from a milkmaid's blisters. He inserted the pus into a cut on the arm of an eight-year-old boy called James Phipps.

A few days later, Jenner then deliberately injected Phipps with smallpox. Phipps became ill but after a few days he made a full recovery. The cowpox vaccine had prevented James from getting smallpox.

In the 1970s smallpox was wiped out from the world as a result of immunisations. No cases have occurred since.

▲ Edward Jenner, the English doctor who discovered the smallpox vaccine.

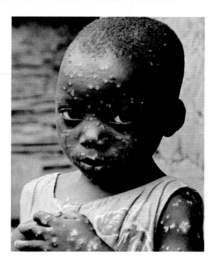
▲ Those who survived smallpox were left badly scarred and often blind.

Key Words
immunisation, vaccine, immune system

Summary Questions

1. 🧪 Copy and complete the sentences below.
   The spread of infectious _____ can be prevented by _____ . They work by introducing _____ that contain dead or inactive _____ into the body.
   (4 marks)

2. 🧪🧪 Explain why dead or inactive microorganisms are used in vaccines.
   (2 marks)

3. 🧪🧪 Suggest three reasons why it is important for scientists like Jenner to study the spread of a disease.
   (3 marks)

4. 🧪🧪🧪 Explain in detail how Jenner discovered the smallpox vaccine.
   (6 marks QWC)

Human experiments
Was Edward Jenner right to test his hypothesis on James Phipps? Discuss the risks and benefits of the immunisation experiment.

Smallpox vaccine
Produce a cartoon strip showing how Jenner developed the smallpox vaccine.

2.2 Vaccines 2

Learning objectives

After this topic you will be able to:
- describe how a person develops immunity
- compare the advantages and disadvantages of receiving a vaccine.

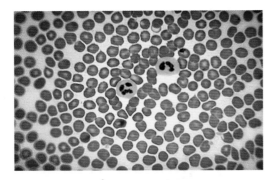

▲ The large white blood cells are responsible for fighting disease. The small cells are red blood cells.

Have you ever heard the word 'immunity'? Doctors and nurses talk about you being **immune** to a disease. This means that you will never get that disease.

What is immunity?

The body has defences to prevent microorganisms entering your body. These include your skin, and the hairs in your nose. Sometimes though, a harmful microorganism, a **pathogen**, does get in. This is detected by your immune system. It will trigger your white blood cells to make **antibodies**. Antibodies are special chemicals that attack and destroy the microorganism. A different type of antibody has to be produced for each type of microorganism.

A Describe what an antibody does.

It takes time for the body to make enough antibodies to destroy pathogens. During this time the microorganisms reproduce rapidly, damaging cells and making poisons that make you ill.

If the same type of pathogen enters your body again, your white blood cells remember it. Antibodies are produced more quickly. The pathogen is then destroyed before it has a chance to make you ill. You then have immunity to the disease.

B Name the cells that make you immune to a disease.

How do immunisations work?

When a vaccine is inserted into the body, white blood cells make antibodies to fight against the dead or inactive microorganism contained in the vaccine. The antibodies will be remembered by your body. If a live version of the pathogen later enters your body, the antibodies will destroy it before it causes disease.

Between 1970 and 2000, the population of the UK increased. During this time, measles immunisations were given to large numbers of the population. Despite a rise in population, the number of measles cases decreased.

▲ Number of cases of measles in the UK between 1970 and 2000.

Immunisations protect you against many life-threatening conditions. However, not everyone chooses to be immunised. This is because they may have concerns about:

- the safety of some vaccines
- possible side effects.

C Suggest one reason why people may choose not to be immunised.

Side effects

Immunisations sometimes have side effects. These can include:

- a temperature
- sickness
- swollen glands
- a small lump at the site of the injection

Side effects can usually be easily treated with a painkiller. Severe reactions are very rare. Before they are widely used vaccines are thoroughly tested to ensure they are safe.

D State two common side effects of immunisation.

MMR and autism

MMR is a vaccine given to young children, which protects against measles, mumps, and rubella. Some people are concerned that the vaccination is linked to autism in children.

The graph shows the number of UK cases of autism across a 15-year period, and the MMR vaccination rate across the same period of time.

Using evidence from the graph, explain why scientists do not believe that autism is linked to the MMR vaccine.

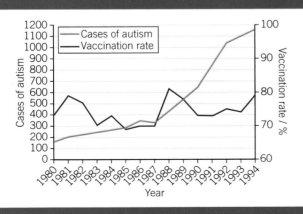

Foul Fact

Bacteria can reproduce very quickly. In ideal conditions they divide into two every 20 minutes. Within a few hours, a few bacteria will have become several million!

Key Words

immune, pathogen, antibody

Summary Questions

1 Copy and complete the sentences below.

When a harmful _____ enters the body, white blood cells make _____ to destroy it. If the same microorganism enters your body again, your _____ _____ _____ remember it. They make the antibodies much more quickly, destroying the microorganism before it causes disease. You have _____ to the disease.

(4 marks)

2 Describe how a vaccine causes immunity.

(4 marks)

3 Compare the advantages and disadvantages of being immunised.

(6 marks QWC)

2.3 Antibiotics 1

Learning objectives

After this topic you will be able to:
- describe the use of antibiotics
- describe how Fleming discovered penicillin.

Have you ever had a very sore throat? Some sore throats are caused by bacteria. Doctors use antibiotics to treat this type of infection.

What are antibiotics?

When you are ill, doctors can prescribe a range of drugs to make you feel better. These medicinal drugs work by preventing, treating, or curing the symptoms of a disease.

A Name three ways that a medicinal drug can work.

Antibiotics work by killing the bacteria that has made you ill. They do not damage the cells in your body, and have no effect on viruses or fungi.

B State what antibiotics do.

There are many different types of antibiotics. One of the most common types is penicillin.

How was penicillin discovered?

During World War 1, Alexander Fleming worked in the battlefield hospitals. He saw large numbers of soldiers dying from wounds infected with bacteria. After the war he carried out research to try and discover chemicals that could be used to kill the bacteria.

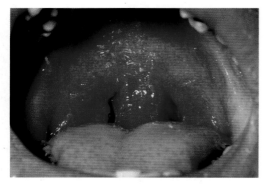

▲ A sore throat caused by a bacterial infection.

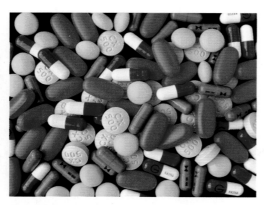

▲ Antibiotics can be taken in tablet or liquid form. For severe infections, antibiotics can be given directly into the blood through a drip.

▲ Alexander Fleming discovered penicillin accidentally.

Fleming carried out his research by growing bacteria on agar plates. In September 1928, he returned to work after a holiday. He had left a number of agar plates with bacteria growing on them stacked up in his laboratory. When he returned he found mould (*Penicillium notatum*) growing on one of them.

He noticed that where the mould was growing, bacteria were killed. Fleming named the substance that killed the bacteria penicillin. Fleming discovered that penicillin killed many types of bacteria, such as those that cause meningitis and scarlet fever.

Fleming's discovery has saved millions of lives, and is recognised as one of the greatest medical developments of all time.

C Name the scientist who discovered the first antibiotic.

Which antibiotic?

There are now a number of different types of antibiotic a doctor can prescribe. Each type kills different species of bacteria.

To discover the correct antibiotic to use, samples of the bacteria can be spread over the surface of an agar plate. Antibiotic discs are then placed on the surface of the agar. These are tiny discs of paper that have been soaked in a particular antibiotic. The agar plates are then placed in an incubator and left for the bacteria to grow.

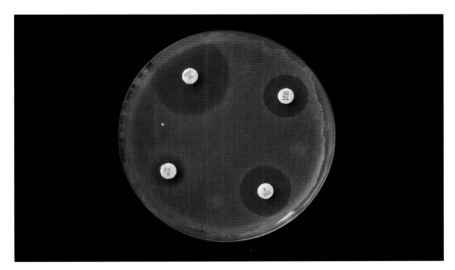

▲ The disc with the largest area free of bacteria is the most effective antibiotic.

This photo of an agar plate shows clear circles around the antibiotic discs. These are the areas where the bacteria have come into contact with the antibiotic, and have been prevented from growing. The larger this area, the more effective the antibiotic is on that type of bacteria.

▲ The mould growing on this orange is *Penicillium notatum*.

Key Words

antibiotic

Summary Questions

1 🧪 Copy and complete the sentences below.

_____ are a type of medicinal _____ . They treat a disease by killing the _____ that caused it. Alexander Fleming discovered the antibiotic _____ . It is made from a type of _____ .

(5 marks)

2 🧪🧪 Colds are caused by viruses. Explain why doctors do not prescribe antibiotics to treat colds.

(2 marks)

3 🧪🧪 Describe how Fleming discovered penicillin.

(4 marks)

4 🧪🧪🧪 Ear infections are caused by a number of different bacteria. Explain in detail how doctors can discover the most effective antibiotic to prescribe to treat the infection.

(6 marks QWC)

2.4 Antibiotics 2

Learning objectives

After this topic you will be able to:
- describe what is meant by antibiotic resistance
- describe some methods for preventing the spread of bacterial infection.

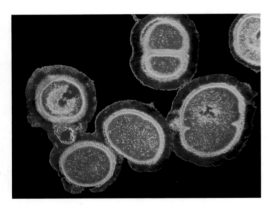

▲ MRSA bacteria are resistant to many antibiotics. They are an example of a superbug.

Preventing MRSA
Design a poster to be placed in hospital waiting rooms that show how the risk of spread of MRSA can be reduced.

Link
You can learn more about how organisms evolve in B2 3.6 Natural selection

Have you heard of 'superbugs'? They are often the cause of infections in hospitals. These infections can be very difficult to treat.

Do antibiotics kill all bacteria?

Since penicillin was first discovered, millions of lives have been saved. Antibiotics have been widely used to treat a range of human and animal conditions. They can even be added to animal feed to prevent infections.

However, some types of bacteria are no longer killed by penicillin. They have become **antibiotic resistant**. New types of antibiotics have been developed but some bacteria are now resistant to these antibiotics too.

A State what it means when a bacterium is antibiotic resistant.

Some types of bacteria are resistant to most types of antibiotic. They are known as **superbugs**. Scientists are worried that these bacteria will become resistant to all antibiotics, leaving no way of treating the infections they cause. An example of a superbug is MRSA.

B State what is meant by a superbug.

Developing resistance

Bacteria reproduce very rapidly. When they multiply their DNA can be damaged or altered. This is known as a **mutation**. This normally results in the bacteria dying but sometimes the mutation can be beneficial to the bacterium. For example, a mutation may cause it to be resistant to an antibiotic.

C State what a mutation is.

Antibiotics should be taken over a period of time to ensure that all the infecting bacteria are killed. Some people stop taking antibiotics too soon because they feel better. This increases the chance of an antibiotic-resistant strain of bacteria developing. The diagram on the next page shows how this happens.

● B3 Chapter 2: Turning points in biology

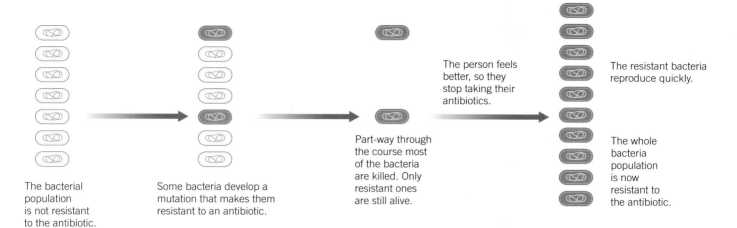

The bacterial population is not resistant to the antibiotic.

Some bacteria develop a mutation that makes them resistant to an antibiotic.

Part-way through the course most of the bacteria are killed. Only resistant ones are still alive.

The person feels better, so they stop taking their antibiotics.

The resistant bacteria reproduce quickly.

The whole bacteria population is now resistant to the antibiotic.

◀ Doctors and nurses rub their hands with antiseptic gel between seeing patients. This helps to prevent the spread of bacteria around a hospital.

How can we stop superbugs developing?

Scientists cannot stop antibiotic-resistant bacteria developing. However, there are a number of things we can all do to reduce the risk of bacteria spreading:

- Thoroughly wash your hands before meals, before preparing food, and after going to the toilet. This removes bacteria from your hands.
- Use antiseptics to clean cuts and grazes. These are chemicals that kill microorganisms but do not damage your skin.
- Clean toilets and kitchens with disinfectants. These are very strong chemicals that kill microorganisms.
- Use sterile medical equipment. Only use plasters and dressings that are sealed. Sterile objects have no microorganisms on them.

D Write down what sterile means.

◀ Surgical equipment is sterilised in an autoclave. This heats the equipment to 120 °C, killing all microorganisms. This reduces the risk of infection from surgery.

Key Words

antibiotic resistant, superbug, mutation

Summary Questions

1 🧪 Copy and complete the sentences below.

Bacteria that cannot be killed by an _____ are antibiotic _____. Bacteria that are resistant to most antibiotics are called a _____. The spread of bacteria can be prevented by good personal hygiene. Medical equipment should be _____. This means that there are no _____ present.

(5 marks)

2 🧪🧪 Describe three ways in which the spread of bacteria can be reduced in hospitals.

(3 marks)

3 🧪🧪🧪 Explain how not completing a prescribed course of antibiotics increases the risk of antibiotic-resistant bacteria developing.

(6 marks QWC)

29

2.5 DNA

Learning objectives

After this topic you will be able to:
- describe the structure of DNA
- describe how scientists worked together to discover the structure of DNA.

Link

You can learn more about DNA and genes in B2 3.5 Inheritance

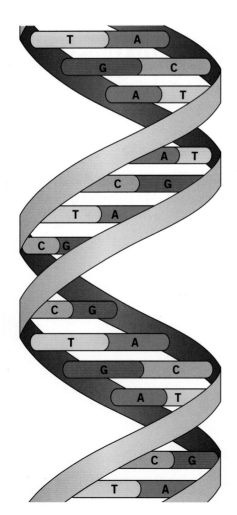

▲ DNA is a double helix.

You know that the nucleus of your cells contains DNA. DNA contains all of the instructions that determine your characteristics.

What does DNA look like?

The chemical **DNA** (deoxyribonucleic acid) contains all the information needed to make an organism. Short sections of DNA are known as genes. Each gene contains the instruction (the code) for a characteristic.

A State what a gene is.

DNA has three main features:

1. It is made up of two strands.
2. The strands are joined together by chemicals called DNA bases.
3. The strands are twisted together to form a double-helix shape.

B Describe three features of a DNA molecule.

What are DNA bases?

DNA contains four different chemical bases. They are normally referred to by the letters A (adenine), T (thymine), C (cytosine), and G (guanine).

The order of the bases is a code for the order of the amino acids, which make a specific protein. The protein determines the function of a cell.

C Name the four DNA bases.

DNA model

Produce a scientific model to demonstrate the structure of DNA.

How did scientists discover DNA?

Many scientists have worked together to discover the structure and function of DNA. Although we have discovered lots of information about DNA, research is still continuing. It is hoped that this will lead to the prevention and cure of many diseases in the future.

The table below shows some of the main steps in the discovery of DNA.

1866	**Certain characteristics are inherited.** Gregor Mendel carries out experiments using peas. He notices that certain characteristics such as height and colour are passed on from parents to their offspring.
1869	**Nuclein is discovered.** Friedrich Miescher discovers an acidic substance in the nucleus of a cell. He calls this substance nuclein. This chemical is now called DNA.
1944	**Genes are passed from one generation to the next.** Oswald Avery transfers the ability to cause disease from one type of bacteria to another. He proves that genes are sections of the DNA molecule.
1950	**DNA base pairs are discovered.** Erwin Chargaff finds out that, even though different organisms have different amounts of DNA, all DNA contains equal quantities of bases, called A, T, C, and G.
1952	**DNA crystals are photographed.** Maurice Wilkins and Rosalind Franklin use X-rays to take an image of DNA crystals.
1953	**Double-helix structure of DNA is identified.** Building on these discoveries, James Watson and Francis Crick publish their description of DNA. They describe it as a double helix – two spirals held together by base pairs.
1953–2000	**Advances in genetics** Individual genes that code for genetically inherited disorders such as cystic fibrosis are discovered. The production of genetically engineered food and animal cloning also begin.
2003	**Human genome project completed** Scientists working across the globe identify around 24 000 genes – the complete set of genes in the human body.

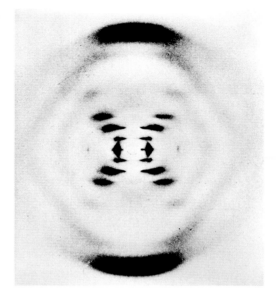

▲ Wilkins and Franklin's famous 'Photo 51' revealed the helical structure of DNA to Watson and Crick. The fuzzy X-shape suggests a helical structure.

Key Words

DNA

Fantastic Fact

The police use a technique called DNA fingerprinting to prove the presence, or absence, of a suspect at a crime scene.

Summary Questions

1. Copy and complete the sentences below.

   DNA is made up of two _____ that are twisted together to form a double _____. The strands are held together by four _____, called A, T, _____, and G.

   (4 marks)

2. Describe how DNA leads to the production of different proteins.

   (2 marks)

3. Explain how scientists worked together to discover the structure and function of DNA.

   (6 marks QWC)

2.6 Charles Darwin

Learning objectives

After this topic you will be able to:
- describe the process of peer review
- describe the evidence that Darwin used to develop his theory of natural selection.

Have you heard of the scientist Charles Darwin? One of the most famous scientists of all time, Darwin developed the theory of evolution.

Darwin's theory

Darwin's theory states that organisms evolve as a result of natural selection. Darwin realised that organisms best suited to their environment are more likely to survive and reproduce, passing on their characteristics to their offspring. Gradually, a species changes over time. We now know that these characteristics are passed on through genes.

A Name the process by which organisms evolve.

How did Darwin come up with his theory?

Darwin was born in 1809. At that time, most people believed that the Earth and all the organisms on it were created by God.

In 1831, Darwin joined Captain Robert FitzRoy's scientific expedition to the Galapagos Islands. Whilst on HMS Beagle, Darwin read Lyell's 'Principles of Geology'. This suggested that fossils were actually evidence of animals that had lived millions of years ago. Modern scientists agree with this.

▲ Charles Darwin, author of 'On the Origin of Species'. Darwin's theory of evolution took over 20 years to develop.

▲ Darwin noticed that finches on different islands had different beaks. The shape of the beak was adapted to the food the finch ate.

Link

You can learn more about evolution in B2 3.6 Natural selection

Darwin noticed that different islands had different types of finch. The birds' beaks and claws were different sizes and shapes. Darwin realised that the size and shape were linked to the type of food available on each island.

B Name the organism that Darwin studied on the Galapagos Islands.

Darwin concluded that if a bird was born with a beak suited to the food available on its island, it would survive for longer. Therefore, it would have more offspring. Over time the population of birds on that island would all have this characteristic. Darwin called this process natural selection.

Another scientist, Alfred Wallace, was working on his own theory of natural selection and evolution. Wallace and Darwin read each other's unpublished work. This is an early example of **peer review**, where a scientist's work is checked by another scientist who works in a similar area of science. Darwin's and Wallace's ideas were so similar that they jointly published the theory of evolution in a scientific paper.

C State what is meant by peer review.

A year later, in 1859, Darwin published his book 'On the Origin of Species'. The book was extremely controversial; the theory of evolution went against the view that God had created all of the life on Earth. As a result of Darwin's theory, people learned that humans were simply a type of animal, and had evolved from apes.

Do people still disagree with Darwin?

Darwin's theory of evolution is now widely accepted, though not by everyone. Evidence for his theory includes:

- the fossil record – which provides evidence that organisms have changed over time
- changes that have been observed in microorganism populations – for example, the development of antibiotic-resistant bacteria
- extinction – species that do not adapt to environmental changes die out.

Natural selection
Imagine you are a newspaper reporter at the time of the publication of Darwin's 'On the Origin of Species'. Write a front-page article about the theory of evolution.

▲ Alfred Wallace gathered his evidence for the theory of evolution from the wildlife of South America and Asia.

Key Words
peer review

Summary Questions

1. Copy and complete the sentences below.
   Charles _____ came up with the theory of _____ by natural _____. Before they published their theory, Darwin and Wallace checked each other's work. This is called _____ _____.
   (4 marks)

2. Describe three pieces of evidence in support of Darwin's theory of natural selection.
   (3 marks)

3. Describe the process of peer review.
   (2 marks)

4. Explain how Darwin's observations of finches in the Galapagos Islands contributed to his theory of natural selection.
   (6 marks QWC)

2.7 Preventing extinction

Learning objectives

After this topic you will be able to:
- describe how animals become extinct
- describe some techniques used to prevent extinction.

▲ A woolly mammoth – this animal became extinct about 4000 years ago.

Link

You can learn more about why species die out in B2 3.7 Extinction

▲ It is estimated that there are fewer than 2000 giant pandas living in the wild.

Several million years ago, dinosaurs roamed the planet. These species are now extinct. Did you know that other species are becoming extinct today?

What does extinct mean?

Extinct means that no organisms of a particular species are alive anywhere in the world. The fossil record shows us that throughout history many species have become extinct. Extinction is still happening today, in many cases as a result of human activity. Humans compete with other organisms for space, food, and water, and are also very successful predators.

A Write down what extinct means.

Can we prevent extinction?

Species that are at risk of extinction are called **endangered species**. This means that there are very few of the species left. An example is the giant panda. Their numbers have been severely reduced by loss of habitat, and by being killed by poachers.

B State why the panda is an endangered species.

There are a number of ways that scientists are trying to prevent extinction. These include:

- **conservation**
- **captive breeding**
- **seed banks**

What is conservation?

Conservation means protecting a natural environment, to ensure that habitats are not lost. Protecting an organism's habitat increases their chance of survival, allowing them to reproduce.

As well as reducing the risk of a particular species becoming extinct, conservation also:

- reduces disruption to food chains and food webs
- makes it possible for medicinal plant species to be discovered.

The UK has over 4000 conservation areas where habitats are protected. These are known as Sites of Special Scientific Interest (SSSI) and cover around 8% of the nation's land.

C State what is meant by conservation.

What is captive breeding?
Captive breeding means breeding animals in human-controlled environments. Scientists working on captive-breeding programmes aim to:
- create a stable, healthy population of a species
- gradually re-introduce the species back into its natural habitat.

D List the aims of captive breeding.

Unfortunately there are also problems associated with captive breeding.
- Maintaining genetic diversity can be difficult. Only a small number of breeding partners are available.
- Organisms born in captivity may not be suitable for release in the wild. For example, predators bred in captivity may not know how to hunt for food.

What are seed banks?
Seed banks are a way of conserving plants. Seeds are carefully stored so that new plants may be grown in the future. A seed bank is an example of a gene bank – a store of genetic material.

The Millennium Seed Bank Project at Kew Gardens is an international project. Its purpose is to provide a back-up against the extinction of plants in the wild by storing seeds for future use. Its large underground frozen vaults preserve over a billion seeds; it is the world's largest collection of seeds.

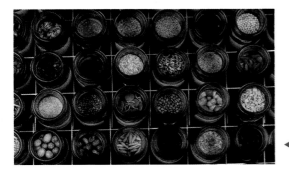

◀ Seeds in the Millennium Seed Bank.

▲ Nearly half of the medicines used by doctors today are based on plant extracts that have come from the rainforests.

Key Words
endangered species, conservation, captive breeding, seed bank

Summary Questions

1 Copy and complete the sentences below.
Scientists are using a number of techniques to try to prevent _____ species becoming _____. These include storing seeds in seed _____, breeding animals in _____, and protecting habitats through _____.
(5 marks)

2 Describe how animals become extinct.
(3 marks)

3 Describe two advantages and two disadvantages of captive-breeding programmes.
(4 marks)

4 Discuss how humans can have a positive or a negative effect on the population of a species.
(6 marks QWC)

Captive-breeding debate
Hold a debate to discuss the advantages and disadvantages of captive-breeding programmes.

B3 Chapter 2 Summary

Key Points

- When a harmful microorganism enters the body, white blood cells make antibodies to destroy it.
- If the same microorganism enters your body again, your white blood cells make antibodies much more quickly, destroying the microorganism before it causes disease. This is known as immunity.
- The spread of infectious diseases can be prevented by immunisations. These work by introducing vaccines, which contain dead or inactive microorganisms, into the body. These trick your immune system into producing antibodies.
- Edward Jenner discovered the first vaccine. It was used to prevent smallpox.
- Antibiotics are medicinal drugs that work by killing bacteria.
- Alexander Fleming discovered the antibiotic penicillin.
- Bacteria that cannot be killed by an antibiotic are antibiotic resistant. This resistance is created by a mutation in their DNA.
- Bacteria that are resistant to many antibiotics are called superbugs. An example is MRSA.
- The spread of bacteria can be prevented by good personal hygiene and the use of disinfectants. Medical equipment should be sterilised before use, to remove all microorganisms.
- DNA (deoxyribonucleic acid) is made up of two strands twisted together to form a double helix. The strands are held together by bases. There are four bases: A, T, C, and G.
- Many scientists worked together to discover the structure and function of DNA.
- Charles Darwin and Alfred Wallace came up with the theory of evolution by natural selection.
- To prevent endangered species becoming extinct, scientists store seeds in seed banks, breed animals in captivity, and protect habitats through conservation.

Key Words

immunisation, vaccine, immune system, immune, pathogen, antibody, antibiotic, antibiotic resistant, superbug, mutation, peer review, endangered species, conservation, captive breeding, seed bank

Big Write

The scientists behind the science
Behind every scientific discovery is a scientist or a team of scientists who have worked together.

Task
Imagine you are working for a magazine. Write a feature article on at least two scientific discoveries from the following list:

- the first vaccine
- the development of the first antibiotic
- the discovery of the structure of DNA
- the development of the theory of evolution

Tips
- Your article should be written in a 'magazine' style. Think carefully about the language you will use, and the way you will present the article.
- Explain any scientific terms used. Your readers may not have much scientific understanding.

End-of-chapter questions

1. Match the scientist to their discovery.

   Darwin — penicillin

   Fleming — smallpox vaccine

   Watson and Crick — theory of evolution by natural selection

   Jenner — structure of DNA

   (4 marks)

2.
   a. Scientists are trying to prevent endangered species becoming extinct. Tick the cells in the table to show what type of organism the technique is used for. **(3 marks)**

Technique	Plants	Animals	Both
captive breeding			
seed bank			
conservation			

   b. Describe the difference between an endangered organism and an extinct organism. **(2 marks)**

   (5 marks)

3. Re-arrange the following sentences to explain how an immunisation works.

   Microorganisms are destroyed before you get ill.

   White blood cells 'remember' the microorganism.

   A vaccine is inserted into the body.

   Antibodies destroy the microorganism.

   White blood cells make antibodies against the dead or inactive microorganism.

   If the live microorganism enters the body, antibodies are made very quickly.

   (6 marks)

4. To work out the most effective antibiotic to prescribe to a patient, a sample containing microorganisms was sent for testing in a laboratory.

   a. Name the type of microorganism an antibiotic is used to kill. **(1 mark)**

   b. State **two** safety precautions that should be followed when working with microorganisms. **(2 marks)**

   c. Before the testing is started, name the process that should be used in the laboratory to ensure all equipment is free from microorganisms. **(1 mark)**

   d. These are the results achieved in the laboratory after using antibiotic discs and bacteria grown on an agar plate.

Antibiotic	Area free of bacteria (cm^2)
A	2
B	4
C	0

   State and explain which antibiotic should be prescribed to the patient. **(3 marks)**

   (7 marks)

5.
   a. Explain how MRSA became a superbug. **(2 marks)**

   b. Explain how hospitals can help reduce the spread of MRSA. **(2 marks)**

   c. Explain how people can help stop antibiotic resistance. **(2 marks)**

   (6 marks)

6. Discuss the importance of the discoveries of vaccines, antibiotics, DNA, and evolution.

   (6 marks QWC)

3.1 Microscopy

Learning objectives
After this topic you will be able to:
- describe the main differences between a light microscope and an electron microscope
- describe how microscopic evidence is used by forensic scientists.

Key Words
forensic science, magnification, resolution

Link
You can learn more about using a light microscope in B1 1.1 Observing cells

Have you ever seen police tape sealing off a crime scene? It is very important that people do not enter a crime scene until forensic scientists have collected any evidence to help them solve the crime.

What is forensic science?
Forensic science is the study of materials and situations that relate to a crime. Forensic scientists gather and study evidence so that it can be used in a court of law.

A Write down a definition of forensic science.

What are microscopes used for?
Many types of evidence that forensic scientists collect are too small to see in detail with the naked eye. Some samples are looked at through light microscopes – this is the type of microscope you use at school.

Some samples are too small even for light microscopes. These are studied using a type of microscope called an electron microscope. These are very powerful microscopes that provide better magnification and resolution.

- **Magnification** is how many times bigger the image appears compared to the object.
- **Resolution** is how clearly the microscope can distinguish two separate points.

B Name two types of microscope that a forensic scientist might use.

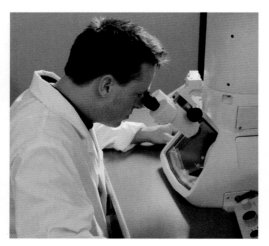

▲ This scientist is looking into an electron microscope. It is much larger and more expensive than a light microscope.

Converting units
A typical cotton fibre is 10 micrometres wide. To understand what this means, you can use the following conversions:

1 cm = 10 mm   1 mm = 1000 micrometres

Convert the following units:

a 8 cm to millimetres

b 6 mm to micrometres

c 3500 micrometres to centimetres

The table below summarises the main differences between light microscopes and electron microscopes.

Microscope	Light	Electron
Radiation source	light	electrons
Typical magnification	up to 1000 ×	up to 1 000 000 ×
Resolution	0.2 μm (2×10^{-7} m)	50 pm (5×10^{-11} m)
Image	colour	black and white
Image of a chloroplast in moss cell		

Uses of microscopes

Microscopes are used to study a range of samples. These include hairs and fibres, paint flecks, pollen grains, and soil. Forensic scientists try to match samples found at a crime scene with those on a suspect's clothing or body.

C Name three types of sample that forensic scientists observe under a microscope.

Pollen

Pollen grains often attach to clothing. Pollen grains differ in size, shape, and surface texture. Pollen samples can provide evidence that a suspect was present at a crime scene.

Clothing fibres

Fibres from clothing look very different under the microscope.

Fibres found at the scene of a crime can be matched to a suspect's clothing, showing that the suspect was present at a crime scene. Also, fibres taken from a suspect's clothing can be matched to a victim's clothes or hair, for example.

D State one difference in appearance between cotton and wool fibres.

Evidence from pollen and clothing fibres can place a suspect at the scene of a crime but this does not always mean that they were there when the crime was committed or that they committed the crime.

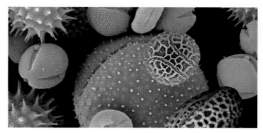
▲ Pollen grains can be useful evidence.

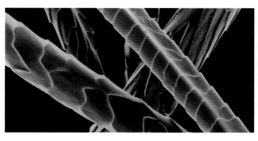
▲ Wool.

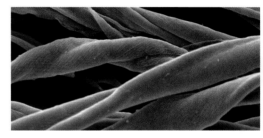
▲ Cotton.

Summary Questions

1 Copy and complete the sentences below.
_____ scientists study evidence so that it can be used in a court of _____. They use a _____ to study samples of fibres. They use two types of microscope – light microscopes and _____ microscopes.
(4 marks)

2 Describe three differences between a light microscope and an electron microscope.
(3 marks)

3 Explain how the use of microscopes can match a suspect to the scene of a crime.
(6 marks QWC)

3.2 Fingerprinting

Learning objectives

After this topic you will be able to:
- describe how fingerprints are formed
- describe how the police use fingerprints to solve crimes.

Fantastic Fact

There is a 1 in 64 billion chance that your fingerprint will match up exactly with someone else's.

Key Words

fingerprint

Have you ever left sticky fingerprints on a table top or on a window? You could claim they were not yours but if you look closely enough you will see that each person's fingerprints are unique.

What is a fingerprint?

The surface of your fingers is covered in tiny ridges. These allow you to grip objects. They form when a baby stretches or bends their fingers in the uterus. The ridges form a pattern that is unique to each individual. **Fingerprints** are an impression of these ridges.

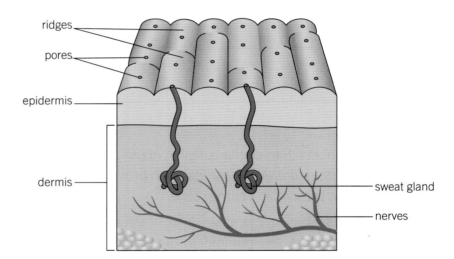

▲ A section of skin.

A State the purpose of ridges in the surface of your fingers.

How are fingerprints made?

Fingerprints can be left in materials such as soil and sand. This is a direct impression of the finger. It provides a mirror image of the ridges and troughs of the finger.

Each ridge contains pores, which are attached to sweat glands under the skin. This is where sweat is released. Oils are left behind after the sweat has evaporated, which forms an impression when you touch a surface. This makes the fingerprints you normally see on glass.

▲ The fingerprints on this glass are created by oils found on the surface of the fingers.

B Describe how a fingerprint can be left on a glass.

How can fingerprints help to solve crimes?

No two fingerprints are the same, not even identical twins' fingerprints. If a person's fingerprint matches one found at the scene of a crime, the person must have been at that location at some point.

Police officers take suspects' fingerprints by pressing each finger in turn on to an ink pad, and then rolling the finger on a piece of paper. The patterns in the fingerprints are then studied and compared to those found at the crime scene.

Finding fingerprints

The most common ways to recover a fingerprint at a crime scene are:

- powder dusting – powders are brushed over the surface of objects. These stick to oils in the fingerprints, making them show up.
- ultraviolet light – oils in the fingerprints glow when ultraviolet light is shone on them.

Matching fingerprints

Scientists look at the arrangement, shape, size, and number of lines in fingerprint patterns to distinguish one from another.

There are three main patterns that are used to identify fingerprints:

- arch – arches slope up and then down, like very narrow mountains
- whorl – whorls form a circular or spiral pattern
- loop – loops begin on one side of the finger, curve around or upward, and end on the other side.

arch   whorl   loop

◀ There are three types of pattern in a fingerprint.

C State the three types of pattern in a fingerprint.

Identifying your own fingerprint

Make your own fingerprint using an ink pad. Identify the patterns in your fingerprint. Does each of your fingers produce the same print?

▲ Taking a fingerprint.

▲ Fingerprints being collected at the site of a break-in.

Summary Questions

1 Copy and complete the sentences below.

Everybody's fingerprints are _____. They are an impression of the _____ present on the surface of your fingers. When _____ evaporates, oils are left behind. This leaves a _____ where a surface is touched.

(4 marks)

2 Describe two different methods that the police can use to collect fingerprints from a crime scene.

(3 marks)

3 Explain how scientists can use fingerprints to prove that a suspect was present at the scene of a crime.

(6 marks QWC)

3.3 DNA fingerprinting

Learning objectives
After this topic you will be able to:
- state what is meant by DNA fingerprinting
- describe some uses of DNA fingerprinting.

Have you ever heard of a DNA fingerprint? Forensic scientists use evidence from DNA fingerprints to match a suspect to the scene of a crime.

What is a DNA fingerprint?
Everybody's DNA is different. The only exception is identical twins; they have the same DNA because they are formed when one fertilised egg splits into two embryos at an early stage of development.

DNA fingerprinting (or DNA profiling) is the analysis of DNA from body samples, to identify individuals. As your DNA is unique it is possible to work out if the DNA has come from you, or from someone else.

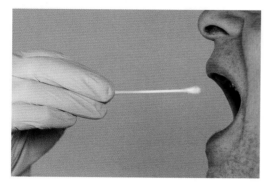

▲ Mouth swabs are used to collect cells from the inside of your cheek.

A State what is meant by DNA fingerprinting.

DNA is found in the nucleus of all body cells (except red blood cells, as they have no nucleus). Therefore, any cell can be used to collect a DNA sample. Most commonly the sample is taken from blood (white blood cells), hair, or cells from the inside of your cheek.

B Name two types of sample that can be used for DNA fingerprinting.

▲ This DNA is being loaded into a well in the gel.

How do you make a DNA fingerprint?
1. DNA is extracted from cells.
2. Enzymes are used to cut the DNA into short fragments.
3. Samples of the DNA are inserted into a well (a small dent) in a gel.
4. An electric current is passed through the gel. This makes the fragments move.
5. The fragments move different distances, depending on their size.
6. The pattern produced is the person's DNA fingerprint. It looks a bit like a barcode.

Presentation
Produce a short presentation for Year 7 students about what a DNA fingerprint is. Describe how this technique is used for solving crimes and in medicine.

Matching DNA fingerprints

To find out whether someone was at the scene of a crime their DNA fingerprint must be compared with samples collected from the crime scene.

The diagram below shows the DNA fingerprint collected at a crime scene. The DNA fingerprints of three suspects are also shown.

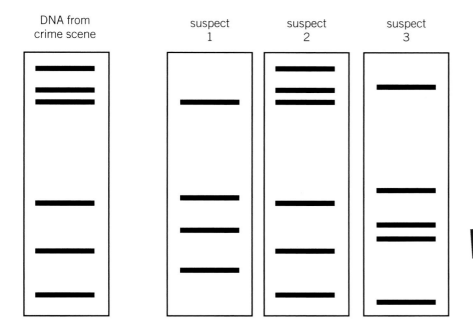

▲ DNA fingerprints of different people.

Can you identify who was at the scene of the crime?

Use your ruler to compare the position of the DNA fragments. You should see that the DNA found at the crime scene is an exact match with the DNA from suspect 2.

Other uses of DNA fingerprinting

DNA fingerprinting is widely used to solve crimes but there are other uses too.

- Maternity and paternity testing – by comparing fragments of DNA you can determine who a child's mother or father is. The child's DNA will be made of fragments of DNA that match the mother, and fragments that match the father.
- Genetic counselling – people wishing to have a child can find out if they are a carrier of a genetically inherited disease. This allows them to make an informed choice about whether or not to have a baby.
- Identification and treatment of genetic disorders – the treatment of some genetic disorders is more successful when they are identified earlier in life.

Key Words

DNA fingerprinting

Summary Questions

1. Copy and complete the sentences below.

   DNA _____ is the analysis of _____ from body samples in order to identify individuals. The DNA is cut into _____ and compared with other samples. Except for _____ twins, each person's DNA is _____. This makes it is possible to work out who the DNA has come from.

   (5 marks)

2. Describe how a DNA fingerprint is made.

   (3 marks)

3. Explain in detail three uses of DNA fingerprinting.

   (6 marks QWC)

3.4 Blood typing

Learning objectives

After this topic you will be able to:
- describe the structure and function of blood components
- describe what is meant by a blood group.

If a violent crime has been committed, blood stains are often left at the scene or on clothing. This forms useful evidence that can help with the identification of a suspect or victim.

What is blood analysis?

Blood analysis is used to help confirm the presence of a suspect at the scene of a crime. It can also be used to identify an unknown victim.

The blood sample is analysed to determine its **blood group**.

A State what is meant by blood analysis.

What is found in blood?

Blood is used to transport materials around your body, and to protect against disease. You have approximately five litres of blood in your body. The main components of blood are:

- red blood cells – to carry oxygen
- white blood cells – to fight disease
- **plasma** – to carry blood cells, digested food, waste (e.g., carbon dioxide), hormones, and antibodies
- **platelets** – fragments of cells that help the blood to clot.

B State the four main components of blood.

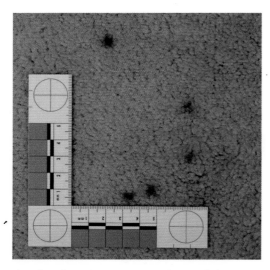

▲ Blood stains can provide useful evidence in solving crimes.

What do blood cells look like?

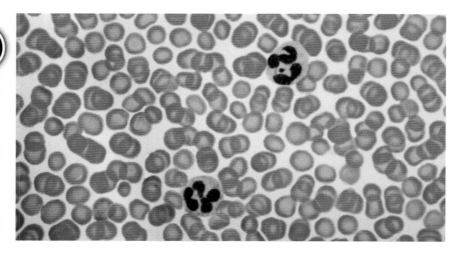

▲ Red blood cells are shown here in red and white blood cells are shown in blue.

Giving blood

When you give blood it is split into its main components. For example, red blood cells are used to 'top-up' blood after major surgery, childbirth, or accidents. Platelets are given to patients being treated for cancers with chemotherapy. Blood transfusions save thousands of lives each year. Write the script for a TV advert to try to persuade people to give blood.

Red blood cells:

- are small
- are a disclike shape and have no nucleus – this increases their surface area for carrying oxygen
- contain haemoglobin (red pigment), which binds to oxygen.

All white blood cells:

- are large
- have a nucleus.

Some white blood cells:

- change shape so they can destroy microorganisms
- produce antibodies.

Plasma:

- is a straw-coloured liquid
- is mainly composed of water (about 90%).

What is a blood group?

The type of blood you have is determined by your genes. There are four main blood groups: A, B, AB, and O.

C Name the four blood groups.

Blood types are determined by antigens (protein molecules) found on the surface of red blood cells:

- Blood group A contains A antigens.
- Blood group B contains B antigens.
- Blood group AB contains both A and B antigens.
- Blood group O contains no antigens.

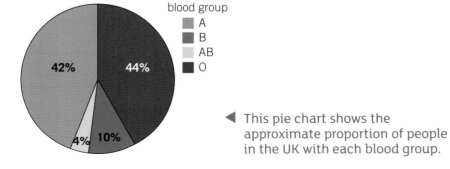

◀ This pie chart shows the approximate proportion of people in the UK with each blood group.

It is important to know your blood group if you need a transfusion. Receiving the wrong group could be life threatening because your immune system will identify the antigens as foreign to your system, and produce antibodies to destroy them.

In 2012, intact samples of red blood cells were identified in the body of Otzi the Iceman, a frozen body discovered in the Italian alps. Otzi died around 3300 BC.

Key Words

blood group, plasma, platelets

◀ Blood transfusions save many lives every day.

Summary Questions

1. 🧪 Match the component of blood to its function.

red blood cells	fight disease
white blood cells	clot the blood
plasma	transport oxygen
platelets	transports blood cells

   (4 marks)

2. 🧪🧪 Explain why it is important for a patient's blood type to be analysed before they receive a blood transfusion.

   (3 marks)

3. 🧪🧪🧪 Compare the structure and function of red blood cells and white blood cells.

   (6 marks QWC)

3.5 Time of death

Learning objectives

After this topic you will be able to:
- describe how time of death can be determined
- describe some of the difficulties in determining time of death.

▲ Scientists use a number of factors to determine the time of death of an animal.

You may have seen flies buzzing around an animal that has been killed on the road. The number and type of insects found can provide evidence of when the animal died.

Finding out the time of death

Scientists use a number of factors to determine the time of death of an animal. These include:
- temperature of the body
- appearance of the body
- insects found on the body.

A State three factors that scientists use to determine the time of death of an animal.

Body temperature

An animal's core body temperature drops by around 1.5 °C per hour from the time of death. This continues until the temperature of the animal matches the temperature of its environment. After this point, scientists have to rely on other factors.

Body appearance

A few hours after death, the body gradually becomes rigid. This process is called **rigor mortis**. The muscles in the body stiffen from a lack of blood and oxygen.

Approximately 20–30 hours later, the rigor mortis disappears. The degree of rigor mortis can be used to estimate the time of death, up to around 48 hours after the animal has died.

B State what is meant by rigor mortis.

From 48 hours after death, the colour of the body can be used to find the time of death. At this point, bacteria begin to breed on the skin, giving the skin a greenish tone.

Approximately 4–7 days after death, skin takes on a marblelike appearance. This occurs as the veins in the body become closer to the surface, making them easier to see. The abdomen starts to inflate because of the gases produced as the body decays.

The Body Farm in Tennessee, US is a research centre that studies how human bodies decay. Bodies are kept in a variety of ways to study the decay process. They have found out that flies and maggots can reduce a body to a skeleton in under two weeks in warm weather.

● B3 Chapter 3: Detection

Insects on a body

The types of insect living on a body, and the stage in their lifecycle, can also be used to determine how long an animal has been dead.

The table below shows the types of insect that are found on a body after death.

Approximate time after death	Example of insects found	
0–3 days	blowflies	
4–7 days	fly larvae (maggots)	
1–2 weeks	cockroaches	
2–4 weeks	mites	

C State the type of insect found on a body after 10 days.

To estimate the time of death accurately, forensic scientists combine information gathered on body temperature, body appearance, and the presence of insects.

Insect identification key

Using the images from this page, produce a key that a forensic scientist could use to help estimate the time of an animal's death.

Key Words
rigor mortis

Summary Questions

1. Copy and complete the sentences below.

   To estimate the time of an animal's death, scientists gather information on the _____ of the body, the appearance of the body, and the presence of any _____ . A few hours after death the body becomes stiff; this is called _____ _____ .

   (3 marks)

2. Describe the condition a body is likely to be in after five days.

   (3 marks)

3. Describe how hot and humid conditions can lead to an inaccurate time of death.

   (3 marks)

4. Explain in detail the techniques that scientists use to determine the time of death of an animal.

   (6 marks QWC)

3.6 Pathology

Learning objectives

After this topic you will be able to:
- describe the role of a pathologist
- describe how dental records can be used to help solve crimes.

Have you heard of a post-mortem? This is an examination of a body carried out by a pathologist (a specialist doctor) to determine the cause of death.

What is a pathologist?

A **pathologist** is a doctor who specialises in understanding the nature and cause of disease. As part of their job, pathologists carry out post-mortem examinations to determine the cause of death.

Forensic pathologists take part in criminal investigations. For example, the police would ask a forensic pathologist to determine whether a suspicious death was an accident or not.

A State what is meant by pathologist.

Pathologists also carry out tests on body samples of living people to determine the cause of disease and illness. For example, they check body tissues for the presence of cancerous cells. Routine tests also performed by pathologists include:

- blood – for example, to check blood–iron levels in detecting anaemia
- urine – for example, to test for sugar in detecting diabetes
- feces – for example, to test for the presence of bacteria in detecting food poisoning.

B State three body samples that pathologists test to identify disease and illness.

▲ Pathologists carry out a post-mortem to determine how a person died.

Key Words

pathologist

Protecting yourself

Working with body samples can be potentially dangerous. Make a list of the safety precautions a pathologist should take so that they do not become contaminated.

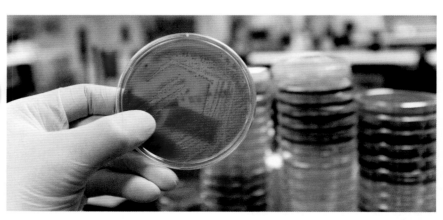

▲ Samples taken from fecal matter have been grown on agar plates to identify the bacteria present.

● B3 Chapter 3: Detection

Identification

If a body cannot be identified by its appearance, dental records are examined. This is a particularly useful technique for a body that has been discovered many years after death. Tooth enamel (the outer layer of teeth) is harder than any other substance in the human body. This means that teeth remain intact long after all other body parts have decayed.

To identify a person from his or her teeth, the teeth are compared to their dental record. Even if only a few teeth are available, a positive identification can normally still be made. The best comparisons come from X-rays the person may have had as part of their dental treatment. If these are not available, dental notes can be used to check if the teeth are the same. For example, a person's dental history may note the presence of fillings or chipped teeth.

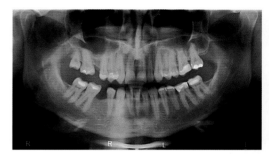
▲ This person has missing teeth, making dental identification straightforward.

C State how a person can be identified by their teeth.

Identifying bite marks

A person's dental records, and impressions of their teeth, can also be used to help solve crimes. Bite marks can be found at the scene of a violent crime. By comparing the features of a bite mark with dental records, it is possible to identify a suspect. This is known as bite-mark analysis.

If a bite mark includes a gap, the biter is probably missing a tooth. Crooked teeth leave crooked impressions, and chipped teeth leave jagged-looking impressions. The depth of a bite mark can also be used to determine how hard a person was bitten.

Victims of fires are often identified using their dental records. Teeth can withstand temperatures of more than 1000 °C.

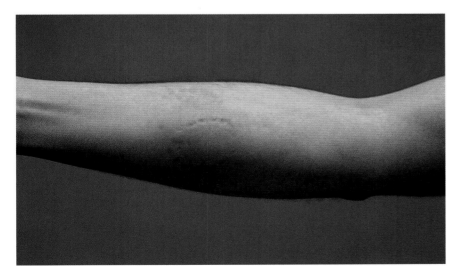
▲ Analysis of a bite mark can be used to solve crimes.

Summary Questions

1. 🧪 Copy and complete the sentences below.
   _____ are doctors who specialise in understanding the nature and cause of disease. They test a range of body _____ such as blood and urine. They also carry out _____ to identify the cause of death.
   (3 marks)

2. 🧪🧪 Describe how dental records can be used to solve crimes.
   (3 marks)

3. 🧪🧪🧪 Explain in detail the role of a pathologist.
   (6 marks QWC)

49

B3 Chapter 3 Summary

Key Points

- Forensic scientists gather and study evidence that can later be used in a court of law.
- Light microscopes and electron microscopes are used to examine objects that are too small to see in detail with the naked eye.
- Electron microscopes provide better magnification and resolution than light microscopes.
- A person's fingerprints are unique. They are an impression of the ridges present on the surface of your fingers. When sweat evaporates, oils are left behind, leaving a fingerprint where a surface was touched.
- DNA fingerprinting is the analysis of DNA from body samples to identify individuals. The DNA is cut into fragments and placed on a gel. When a current is passed through the gel the fragments move, creating a unique pattern.
- The four main blood groups are A, B, AB, and O. Your blood group is determined by antigens (protein molecules) found on the surface of your red blood cells.
- Blood contains plasma, platelets, red blood cells, and white blood cells.
- To estimate the time of an animal's death, scientists gather information on body temperature, body appearance, and the presence of any insects.
- Pathologists are doctors who specialise in understanding the nature and cause of disease. They test a range of body samples such as blood and urine. They also carry out post-mortems to identify the cause of death.
- Dental records can be used to identify bodies that are otherwise hard to identify.

Key Words

forensic science, magnification, resolution, fingerprint, DNA fingerprinting, blood group, platelets, plasma, rigor mortis, pathologist

Case study

To catch a thief
There has been a break-in. The thief cut himself when breaking the window.

Task
Identify different possible sources of evidence in the crime scene. Produce a factsheet to explain how each source can be used in solving the crime.

Tips
- Include an annotated image of the crime scene, showing where forensic scientists should gather samples.
- Explain how each sample should be analysed.
- Explain how the results can be used to solve the crime.

End-of-chapter questions

1. 🧪 Complete the sentences below about blood.

   _____ blood cells carry oxygen around the body.

   _____ blood cells fight disease.

   _____ are used to clot the blood.

   A, AB, B, and O are the four main blood _____.

   (4 marks)

2. 🧪 Forensic scientists often use microscopes to look at evidence they have collected.
   a. State the purpose of a microscope. *(1 mark)*
   b. Name the two types of microscope a forensic scientist might use. *(2 marks)*
   c. Name **two** pieces of evidence a forensic scientist might look at using a microscope. *(2 marks)*

   (5 marks)

3. 🧪🧪 Fingerprints are often left at the scene of a crime.
   a. Describe how forensic scientists identify fingerprints at a crime scene. *(2 marks)*
   b. Suggest **one** advantage of using fingerprints, rather than DNA fingerprints, to help solve crimes. *(1 mark)*
   c. Explain how police take fingerprints from suspects. *(3 marks)*

   (6 marks)

4. 🧪🧪 Forensic scientists collected the following pieces of evidence from a crime scene.
   - a fingerprint on a window pane
   - a sample of blood from a bloodstain
   - a strand of hair

   Explain how each of these pieces of evidence can be used to link a suspect to a crime.

   (4 marks)

5. 🧪🧪🧪 Below is a DNA fingerprint. There are three suspects.

   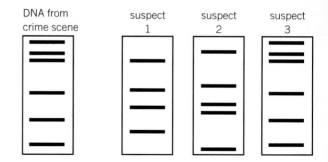

   a. Suggest the source of DNA that may have been found at the crime scene. *(1 mark)*
   b. State and explain which suspect's DNA matches the DNA found at the crime scene. *(2 marks)*
   c. Describe how enzymes are used in the production of a DNA fingerprint. *(1 mark)*
   d. Explain the role of electricity in the production of a DNA fingerprint. *(2 marks)*

6. 🧪🧪🧪 Explain how the processes of microscopy, fingerprinting, DNA fingerprinting, blood typing, and time-of-death estimates can be used to solve a violent crime.

   (6 marks QWC)

Chemistry 3

In this unit you will learn about nanoparticles. Nanoparticles make exciting new materials, with properties that make them perfect for strengthening sports equipment, protecting electronic devices, and treating disease. You will also learn about vital turning points in chemistry. How did scientists find out what's in an atom, and what fossils tell us about the history of life on Earth? You will also discover how chemistry helps to solve crimes.

You already know

- The properties of a substance are determined by the structure and behaviour of its particles.
- The properties of a substance determine its uses.
- Polymers have long particles.
- All substances are made up of atoms.
- Every element has its own type of atom.
- All the elements are listed in the Periodic Table.
- The pH scale shows whether a solution is acidic, alkaline, or neutral.
- Techniques for separating mixtures include filtration and chromatography.

Q Give the pH of a neutral solution, and the pH ranges of acidic and alkaline solutions.

BIG Questions

- What gives nanoparticles their properties, and how do these properties make them useful?
- How did chemists find out what atoms are made of?
- How does chemistry help solve crimes?

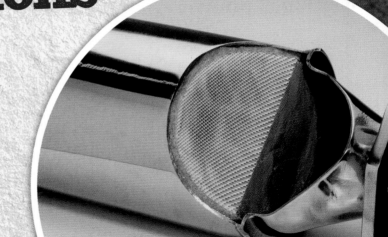

Picture Puzzler
Key Words

Can you solve this Picture Puzzler?

The first letter of each of these images spells out a science word that you will come across in this unit.

Picture Puzzler
Close Up

Can you tell what this zoomed-in picture is?

Clue: It's millions of years old.

Making connections

In **C3** you will learn more about fossils and human ancestors.

In **B3** you will learn about advances in medicine.

In **P3** you will learn how the Universe began.

1.1 Nanoparticles

Learning objectives

After this lesson you will be able to:
- explain what nanoparticles are
- describe the properties of nanoparticles.

Bandages that kill bacteria. Lightweight bullet-proof vests. Materials that capture carbon dioxide. What do these things have in common? They all include nanoparticles, or might in future.

All over the world, scientists are studying nanoparticles. They use what they learn to create exciting new materials.

A List three things that include nanoparticles, or might in future.

What are nanoparticles?

Nanoparticles are tiny pieces of a substance. A nanoparticle is made up of just a few hundred atoms. The diameter of a nanoparticle is between 1 nanometre and 100 nanometres. You cannot see it with a normal microscope.

A **nanometre** is a unit of length. Its symbol is nm. One nanometre is one billionth of a metre, or 0.000 000 001 m. The length of 1 nm compared to 1 m is the same as the diameter of a marble compared to the diameter of the Earth.

When we talk about nanoparticles, "particle" means a piece of a substance.

B State what a nanoparticle is.

▲ Sebastien Zamith is studying nanoparticles to discover how water droplets form in clouds.

Why are nanoparticles special?

A substance that exists as nanoparticles has different properties to normal-sized pieces of the same substance. Why is this?

Imagine a gold ring. It has a mass of 4 g. Most of its atoms are inside the gold. Very few are on the surface.

Now imagine you have 4 g of gold nanoparticles. There is the same number of atoms in the ring and in the nanoparticles. But there are many more atoms on the surface of the nanoparticles.

The huge number of surface atoms gives a substance special properties when it is in the form of nanoparticles.

Normal-sized pieces of gold are yellow in colour. If you mix nanoparticles with water, the water looks reddish in colour. In its normal form, gold cannot mix with water.

▲ A gold ring.

C3 Chapter 1: New technology

◀ The bottle contains gold nanoparticles mixed with water.

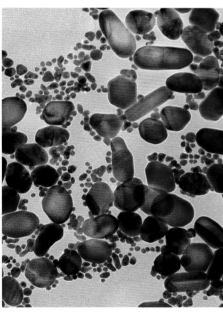

◀ Gold nanoparticles. The picture was taken with a scanning electron microscope, and then coloured. Magnified × 215 000.

C State which has the greater number of atoms on its surface, 1 g of a gold ring or 1 g of gold nanoparticles.

How small?
A human hair has a diameter of between 20 and 100 μm. Calculate the mean diameter of a human hair. Then calculate how many nanoparticles of diameter 30 nm fit across an average human hair.

1 nm = 0.000 000 001 m and 1 μm = 0.000 001 m

Fantastic Fact
Some socks contain silver nanoparticles to kill bacteria and help stop feet smelling.

Key Words
nanoparticle, nanometre

Summary Questions

1. Copy the sentences below, choosing the correct bold words.

   A nanoparticle is a tiny piece of a substance. It is made up a few **hundred/million** atoms. The diameter of a nanoparticle is between 1 and **100/1000** nanometres. One nanometre is **0.000 000 001/ 0.000 001** metres. A substance has **the same/ different** properties when it exists as nanoparticles compared to when it exists in normal-sized pieces. This is because the nanoparticles have a **smaller/ bigger** surface area.

   (5 marks)

2. A scientist has 1 g of silver nanoparticles, and a 1 g lump of silver. Which has the greater number of surface atoms? Explain your answer.

   (2 marks)

3. Compare the properties of gold nanoparticles with those of gold in normal-sized pieces. Explain any differences.

   (6 marks QWC)

1.2 Using nanoparticles

Learning objectives

After this lesson you will be able to:
- explain how the properties of nanoparticles make them suitable for their uses.

Imagine a phone that survives being dropped in the toilet. Imagine clothes that never get dirty. Imagine sports equipment that never breaks. Thanks to nanoparticles, these dreams may soon come true.

Substances in the form of nanoparticles are already useful. Scientists keep thinking of new ways to use their special properties.

What are carbon nanotubes?

Scientists first made **carbon nanotubes** in the 1990s. Carbon nanotubes are cylinders of carbon atoms. Their walls can be just one atom thick. The diameter of a carbon nanotube is about 1 nm.

▲ Nanoparticles may soon mean that your phone could survive being dropped in the toilet.

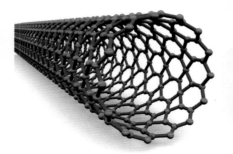

▲ A model showing the atoms in one nanotube.

▲ A cluster of nanotubes.

A State what a carbon nanotube is.

Carbon nanotubes have amazing properties. They are some of the strongest and stiffest materials that exist. They have low densities. These properties make them ideal for many uses, including:
- strengthening tennis-racquet frames
- strengthening bicycle parts
- making lightweight bullet-proof vests.

▲ Composite materials that include carbon nanotubes are used in wind turbines.

C3 Chapter 1: New technology

Mixing carbon nanotubes with polymers makes strong composite materials. These materials are used in wind turbines.

Carbon nanotubes are excellent conductors of heat. Scientists think they could be used in future to heat aircraft wings to stop them going icy. The table shows how well carbon nanotubes conduct heat, compared to other materials.

Material	Thermal conductivity (W/m/K) (how well the material conducts heat)
brick	1.4
carbon nanotubes (along the tube)	3500
carbon nanotubes (across the tube)	1.52
copper	401

B Write down three properties of carbon nanotubes.

Can nanoparticles protect phones and fabrics?

In 2013, scientists announced an exciting invention. They had created a nanoparticle coating for phones and other electronic devices. The coating means there is nowhere for liquids to get in.

The coating consists of polymer nanoparticles. It is 1000 times thinner than a human hair. You cannot see it, and it does not affect how the phone works. When a raindrop (or spilt coffee) falls on the phone, it does not spread out. It stays as a drop, and rolls away.

Other groups of scientists are experimenting with clothing. Can they add nanoparticle coatings to clothes to stop them getting dirty?

C State one use of a nanoparticle coating.

Nanoparticles in the news
Write an exciting news article about one of the new uses of nanoparticles described on these two pages. Start by describing the innovation. Then use simple language to explain why the properties of the nanoparticle material make it perfect for this use.

Fantastic Fact
The longest carbon nanotube ever made was 18.5 cm long.

Key Words
carbon nanotube

Summary Questions

1. Copy and complete the sentences below.
   Carbon nanotubes are _____ of carbon atoms. Their walls are often one _____ thick. The diameter of a carbon nanotube is about one _____. Carbon nanotubes are strong, _____, and they are good conductors of _____.
   (5 marks)

2. Carbon nanotubes have been used to make ice-hockey sticks and baseball bats. Identify the property of carbon-nanotube materials that makes them suitable for these purposes.
   (1 mark)

3. Use the data in the table on this page to write a few sentences to compare the thermal conductivity of the materials in the table. Include an explanation of the term *thermal conductivity* in your answer.
   (6 marks QWC)

1.3 Nanoparticles in medicine

Learning objectives

After this lesson you will be able to:
- describe how nanoparticles are used in medical treatments.

Which of these medical innovations use nanoparticles?

- cancer drugs with no side effects
- bandages that detect infections
- magnetic treatments that destroy tumours

All these innovations use nanoparticles. Scientists think creatively to come up with ideas for new treatments. They test out their ideas by collecting evidence. They tell others about their discoveries. The use of nanoparticles to treat disease is called **nanomedicine**.

A State what is meant by nanomedicine.

How do nanoparticles improve bandages?

Scientists are working on a new type of bandage. The bandage detects and treats burns that get infected by harmful bacteria. Scientists hope that it will prevent deaths caused by infections.

The bandage includes nanoparticles arranged in spheres. Inside each sphere is some dye and an antibiotic.

A nurse bandages a burn wound on a patient. Later, bacteria get into the wound. The bacteria give out poisons. The poisons break into the spheres of nanoparticles. As a sphere breaks, its dye comes out. The dye stains the bandage. The nurse knows that the wound is infected. Next the spheres release their antibiotics, which destroy the bacteria.

Scientists expect the bandages to have many advantages, including:
- detecting and treating infections immediately
- not needing to remove bandages – removing them can be painful, and slow down healing.

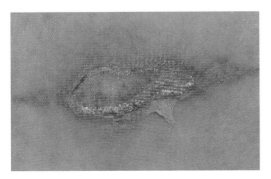

▲ An infected burn.

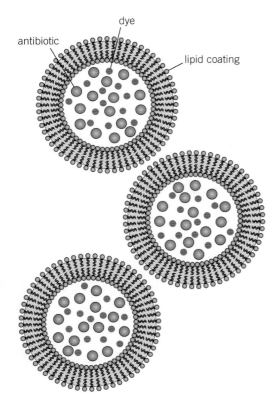

▲ Spheres of nanoparticles like these are used in bandages.

B Describe two advantages of using nanoparticles in bandages.

How can nanoparticles deliver drugs to cancer cells?

Some cancer drugs have very unpleasant side effects. How can we prevent these side effects? Can we use nanoparticles to deliver drugs to cancer cells?

C3 Chapter 1: New technology

Scientists are creating nanoparticles with cancer-fighting drugs inside them. The nanoparticles travel through the body. They reach a cancer cell. The cancer cell recognises the nanoparticles, and absorbs them. Inside the cell, the nanoparticles break up. They release their medicine.

This type of drug-delivery system has many advantages. The drug only gets into cancer cells. This reduces side effects, and does not waste medicine.

Your treatment explained...
Choose one of the treatments described on these two pages. Write a leaflet for a patient who is about to have the treatment. Include a description of what will happen to the patient, and an explanation of how it works.

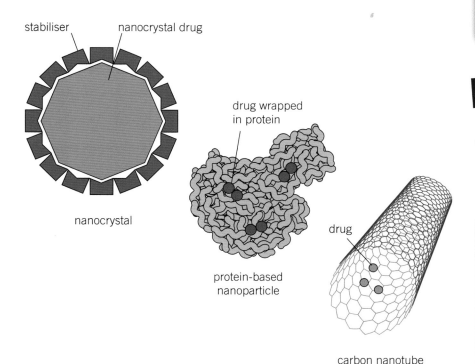

▲ Three types of nanoparticle that deliver drugs.

c Describe how nanoparticles deliver drugs to cancer cells.

How do magnetic nanoparticles destroy tumours?

Scientists are working on other cancer treatments. One possible treatment involves injecting iron oxide nanoparticles into a tumour. The nanoparticles are magnetic.

Doctors will apply a magnetic field near the tumour, and keep changing its direction. This heats the nanoparticles and warms up the tumour. At 5 °C above normal body temperature, tumour cells die. Healthy cells nearby are not harmed.

Scientists are testing this technique on mice with lung cancer.

Key Words
nanomedicine

Summary Questions

1. Copy and complete the sentences below.
   The use of nanoparticles to treat disease is called _____. Scientists are developing many new treatments based on nanoparticles, including bandages that _____ and _____ infection, nanoparticles that _____ drugs to cancer cells, and magnetic nanoparticles that _____ and _____ tumours when an alternating magnetic field is applied.
   (6 marks)

2. Describe how magnetic nanoparticles might one day destroy tumours.
   (3 marks)

3. Make a poster to explain in detail six uses of nanoparticles.
   (6 marks)

1.4 Nanoparticle safety

Learning objectives

After this lesson you will be able to:
- describe an example of how scientists are investigating nanoparticle safety.

▲ Sunscreen with normal zinc oxide is white.

Scientists are finding more and more uses for nanoparticles. But are they safe? Do they cause more health problems than they solve? Scientists are investigating.

Investigating sunscreen

Some sunscreens contain zinc oxide nanoparticles. Zinc oxide stops harmful ultraviolet light reaching the skin. Normal zinc oxide is white. It can be visible on your skin, so is not popular. Zinc oxide nanoparticles have diameters of between 20 and 30 nm. They are invisible.

A Write down the range of diameters of zinc oxide nanoparticles.

Scientists in Australia read about research that others had done. They learnt that nanoparticles could be damaging if they get into cells in the human body. The Australians asked a question about **safety** – *Can zinc oxide nanoparticles get through the skin and into the blood?* They devised an investigation to find out.

The scientists made two types of sunscreen.

- Sunscreen A contained zinc oxide nanoparticles.
- Sunscreen B had bigger pieces of zinc oxide.

B State the question the scientists were investigating.

The scientists asked for volunteers. They took blood and urine samples from the volunteers. They measured the concentration of zinc in their blood.

Half of the volunteers applied sunscreen A twice a day. Half of the volunteers applied sunscreen B twice a day.

After a few days, the scientists took blood and urine samples from the volunteers. There was extra zinc in all the blood samples, but the amounts were tiny. It did not seem to matter which sunscreen the volunteers had used.

The scientists could not detect which form the zinc was in. Was it zinc oxide nanoparticles, which might damage cells? Or was it normal dissolved zinc, which is vital for health? The scientists want to do more research to find out.

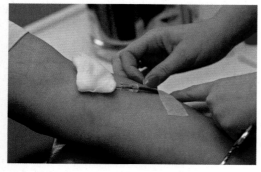

▲ Taking a blood sample.

The Australian scientists considered the benefits and risks of using sunscreen. They suggest that people should continue to use nanoparticle-containing sunscreen. Not using sunscreen can lead to cancers.

Investigating the effects of diesel exhaust fumes

Scientists have been investigating the effect on health of nanoparticles in diesel exhaust fumes.

They did experiments with mice. They bred some mice with high cholesterol. They divided the mice into two groups.

- Group A was exposed to traffic pollution, which includes diesel exhaust fumes.
- Group B breathed clean air.

Five weeks later, the scientists examined arteries in the mice. The arteries of the mice in group A had 55% more fat than the arteries of the mice in group B.

The scientists concluded that diesel exhaust fumes might cause a build-up of fatty deposits in arteries. These can lead to heart disease.

C Describe what the scientists found in their investigation of diesel exhaust fumes.

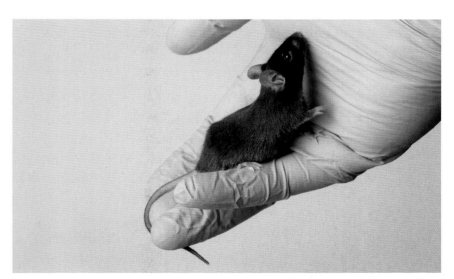

▲ A laboratory mouse.

Spotting variables

With a partner, identify the dependent and independent variables in the sunscreen investigation. Which other variables do you think the scientists would need to consider in their investigation?

▲ Exhaust fumes can be harmful to health.

Key Words

safety

Summary Questions

1. Suggest why scientists are investigating nanoparticle safety.
   (1 mark)

2. a Write down the independent and dependent variables in the mouse experiment.
   (2 marks)

   b List two other variables that scientists would need to consider in the mouse experiment.
   (2 marks)

3. Describe how scientists have investigated the impact of nanoparticles in diesel exhaust fumes on human health.
   (3 marks)

4. Design an experiment to investigate whether zinc oxide nanoparticles from sunscreen get into mouse urine. Identify the variables and consider how you would make sure the experiment is fair.
   (6 marks)

1.5 Cars: pros and cons

Learning objectives
After this lesson you will be able to:
- explain how combustion reactions in car engines produce exhaust gases
- describe some advantages and disadvantages of cars.

Link
You can learn more about greenhouse gases in C2 4.6 Climate change

Key Words
hydrocarbon, particulate

Imagine a world without cars. Would it be a better or worse place to live? Cars are convenient, making it easier for you to get wherever you want, whenever you want.

But cars are not all good news. In 2011 nearly 2000 people died in road accidents in the UK. Car drivers tend to walk less than non-car drivers, and are more likely to become obese. And, of course, cars cause pollution.

What's in car exhaust fumes?

Carbon dioxide
Petrol and diesel are mixtures of compounds. Most of the compounds are **hydrocarbons**. This means they are made up of atoms of carbon and hydrogen only. Octane is one hydrocarbon in petrol. When it burns, it reacts with oxygen to make carbon dioxide and water:

$$\text{octane} + \text{oxygen} \rightarrow \text{carbon dioxide} + \text{water}$$
$$2C_8H_{18} + 25O_2 \rightarrow 16CO_2 + 18H_2O$$

A Name the two products of the combustion reaction of octane.

On a cold day, you can often see water dripping out of car exhaust pipes. Carbon dioxide gas escapes to the atmosphere. It is a greenhouse gas, and contributes to climate change. Scientists estimate that carbon dioxide from cars accounts for 13% of the UK's total carbon dioxide emissions.

Oxides of nitrogen
Car engines get very hot. At high temperatures, nitrogen and oxygen from the air react together. In petrol engines, the main product of this reaction is nitrogen monoxide:

$$\text{nitrogen} + \text{oxygen} \rightarrow \text{nitrogen monoxide}$$
$$N_2(g) + O_2(g) \rightarrow 2NO(g)$$

In diesel engines, the main product is nitrogen dioxide:

$$\text{nitrogen} + \text{oxygen} \rightarrow \text{nitrogen dioxide}$$
$$N_2(g) + 2O_2(g) \rightarrow 2NO_2(g)$$

Fantastic Fact
On average, car drivers walk for one hour less per week than people without cars. Over 10 years, car drivers could gain 13 kg in body mass compared to non-drivers.

C3 Chapter 1: New technology

nitrogen molecule   oxygen molecule   nitrogen dioxide molecules

▲ When nitrogen reacts with oxygen to make nitrogen dioxide, atoms are re-arranged.

If you have asthma, breathing in nitrogen dioxide can make it worse. Nitrogen dioxide in the air dissolves in rainwater to make an acidic solution. This causes acid rain. Acid rain damages trees and lake life, as well as limestone buildings.

B Name the oxides of nitrogen produced when nitrogen and oxygen react together in petrol and diesel engines.

Particulates

Burning diesel also makes another type of product: **particulates**. Particulates are tiny bits of solid, about 100 nm across. They mix with the air. Particulates are easy to breathe in because they are so small. They quickly reach the lungs. They make asthma and chest infections worse.

A group of scientists studied data on the causes of heart attacks. They found that air pollution, including particulates, can lead to heart attacks.

C Identify three health problems caused by diesel particulates.

▲ Acid rain damages trees.

Calculating CO_2

In 1975 the mass of carbon dioxide emitted by UK cars was 46 million tonnes. The mass of carbon dioxide increased by 57% between 1975 and 2005. Calculate the mass of carbon dioxide emitted by UK cars in 2005.

Summary Questions

1. Copy the sentences below, choosing the correct bold words.

   Most cars burn petrol or **diesel/oil**. These fuels are mixtures of **acids/hydrocarbons**. Their molecules consist of atoms of hydrogen and **oxygen/carbon** only. When they burn, they make two products — carbon dioxide and **water/oxygen**.
   (4 marks)

2. One of the hydrocarbons in petrol is heptane, C_7H_{16}. Write a word equation for its combustion reaction.
   (4 marks)

3. Evaluate the advantages and disadvantages of cars.
   (6 marks QWC)

1.6 New fuels

Learning objectives

After this lesson you will be able to:
- describe the advantages and disadvantages of new vehicle fuels.

Petrol and diesel are fossil fuels. They are separated from crude oil. Will they last forever?

Crude oil is non-renewable. It is used up more quickly than it is replaced. And, of course, burning petrol and diesel produces harmful gases. This is why scientists are developing replacement fuels.

A State what is meant by non-renewable.

What are the pros and cons of hydrogen fuel?

There are two ways of using hydrogen fuel. Some cars burn hydrogen in their engines instead of petrol. Another type of car has a **hydrogen fuel cell**. Hydrogen flows into the cell from a tank. It reacts with oxygen from the air and generates electricity. The electricity powers a motor. This makes the wheels turn.

Hydrogen fuel makes one harmless waste product, water. But it does have disadvantages:

- Hydrogen is difficult to store, since mixtures of hydrogen and oxygen are explosive.
- There are few hydrogen filling stations.

Hydrogen does not occur naturally on Earth. It is manufactured from methane in two steps. The process makes waste carbon dioxide.

Step 1: methane + water → carbon monoxide + hydrogen
$$CH_4 + H_2O \rightarrow CO + 3H_2$$

Step 2: carbon monoxide + water → carbon dioxide + hydrogen
$$CO + H_2O \rightarrow CO_2 + H_2$$

Crude oil is extracted from oil wells under the ground or sea.

B Describe one advantage and one disadvantage of hydrogen fuel.

Hydrogen fuels this London bus.

Hydrogen fuels this Bristol ferry.

Fantastic Fact

Swedish scientists have discovered how to make methane from dead cow waste. The methane fuels trains.

Key Words

hydrogen fuel cell, biofuel, renewable, carbon neutral

C3 Chapter 1: New technology

What are biofuels?

If you lived in Brazil, you would travel in vehicles fuelled by **biofuels**. Biofuels are made from plants, or animal waste. They are **renewable**, meaning they are easily replaced.

▲ Brazil makes ethanol from sugar cane. Ethanol burns in car engines.

▲ Oilseed rape seeds contain oil. The oil makes biodiesel.

▲ A few vehicles burn waste cooking oil.

All biofuels are mixtures of compounds. The compounds consist mainly of carbon, hydrogen, and oxygen atoms. They burn to make carbon dioxide and water. Carbon dioxide is a greenhouse gas.

Some people say that biofuels are **carbon neutral**. This means that the plants they are made from remove the same amount of carbon dioxide from the atmosphere when they grow as the fuels put into the atmosphere when they burn.

However, farmers use fossil-fuel-burning tractors when growing crops. They add fertilisers. This means that biofuels are not really carbon neutral.

c State what a biofuel is.

Fuel chart

The table shows the energy released on burning different fuels. Plot the data on a suitable graph or chart to show how the values compare.

Fuel	Energy released on burning (kJ/g)
diesel	45
hydrogen	143
ethanol	30
sunflower oil	38
peanut oil	40
rapeseed oil	37

Summary Questions

1. Copy and complete the sentences below.
   Petrol and diesel are _____ fuels. They are non-_____. Some vehicles use hydrogen fuel. This produces electricity when it reacts with oxygen in a fuel _____. There is one waste product, _____. Fuels produced from _____ or animal _____ are called _____.
   (7 marks)

2. Use the data in the table on this page to calculate the amount of energy transferred on burning 5 kg of sunflower oil.
   (2 marks)

3. Describe the advantages and disadvantages of biofuels.
   (2 marks)

4. Some people say that alternative fuels are carbon neutral. Discuss whether or not you agree with this statement.
   (6 marks QWC)

1.7 Cleaning up exhausts

Learning objectives

After this lesson you will be able to:
- explain using word equations how catalytic converters clean up exhaust gases.

Car exhaust fumes were much more dangerous 50 years ago. What has made them safer?

A modern car has a **catalytic converter** between its engine and exhaust pipe. This converts harmful substances made in the engine to less harmful ones.

▲ A catalytic converter.

A State the purpose of a catalytic converter.

Waste substances

Petrol is a mixture of hydrocarbons. Hydrocarbons burn to make carbon dioxide and water. These are not the only products of combustion. Small amounts of carbon monoxide and nitrogen monoxide are also made.

Some hydrocarbon molecules do not burn. They mix with the combustion products.

It is not safe to release large amounts of these substances into the air.
- Carbon monoxide is poisonous.
- Nitrogen monoxide is a greenhouse gas.
- Some hydrocarbons increase the chance of getting cancer.

nitrogen monoxide   carbon monoxide

unburned hydrocarbons, for example, octane

▲ Petrol engines produce these harmful substances.

B Name three harmful substances made in petrol engines.

How do catalytic converters work?

Reactions to remove harmful substances

Carbon monoxide and hydrocarbons react with oxygen. The reactions make less harmful substances:

carbon monoxide + oxygen → carbon dioxide

$$2CO + O_2 \rightarrow 2CO_2$$

octane + oxygen → carbon dioxide + water

$$2C_8H_{18} + 25O_2 \rightarrow 16CO_2 + 18H_2O$$

On their own, these reactions are very slow. Catalysts in catalytic converters speed them up.

Inside a catalytic converter

Many catalytic converters are made from a ceramic, with a honeycomb structure. This has a big surface area. There is plenty of space for reactions.

Catalyst calculations

Imagine that a company extracted 15 g of platinum, and no other substances, from 3 tonnes of rock. Calculate the mass of waste material produced. 1 tonne = 1000 kg.

C3 Chapter 1: New technology

C State why catalytic converters have a large surface area.

On the surface of the honeycomb is the catalyst. This may include platinum, rhodium, and palladium.

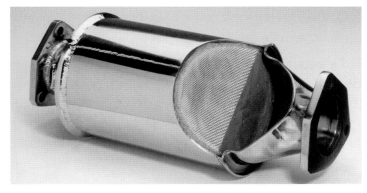

▲ The honeycomb structure of a catalytic converter.

Oxidation reactions take place on the surface of the catalyst:
- Carbon monoxide makes carbon dioxide.
- Unburned hydrocarbons make carbon dioxide and water.

At the same time, nitrogen monoxide decomposes on the catalyst surface:

nitrogen monoxide → nitrogen + oxygen

$$2NO \rightarrow N_2 + O_2$$

D Name three metals in catalytic converters.

Where do the catalysts come from?

Platinum and palladium are unreactive. They are near the bottom of the reactivity series. This explains why they can occur naturally as elements.

Some platinum comes from Canada, where it exists as platinum arsenide ($PtAs_2$). This is mixed with other chemicals in an ore. There is 5 g of platinum in one tonne of rock.

Some metals used in catalytic converters are recycled from old catalytic converters. Every year, around 2 tonnes of rhodium are obtained from this source. About 16 tonnes of rhodium are extracted from rock each year.

Key Words
catalytic converter

▲ Platinum.

Summary Questions

1. 🧪 Copy and complete the sentences below, choosing the correct bold words.

   The main products of hydrocarbon combustion are **carbon dioxide/nitrogen dioxide** and water. Reactions in car engines also make carbon **monoxide/trioxide** and nitrogen **monoxide/trioxide**. Catalytic converters remove the unwanted carbon compound in **an oxidation/a decomposition** reaction.
   (4 marks)

2. 🧪🧪 Use data on this page to calculate:
   a. the total mass of rhodium obtained each year *(1 mark)*
   b. the fraction and the percentage of this mass that is recycled rhodium. *(4 marks)*

3. 🧪🧪 Write word equations to summarise the reactions that occur in the catalytic converter.
   (3 marks)

4. 🧪🧪🧪 Draw a large diagram of a catalytic converter. Add labels and notes to your diagram to show how it removes harmful substances from exhaust gases. Add balanced formula equations to your diagram and include state symbols.
   (6 marks)

1.8 Hybrid electric cars

Learning objectives

After this lesson you will be able to:
- explain why hybrid electric cars use less fuel than cars fuelled by petrol alone
- compare the advantages and disadvantages of different types of car.

Have you ever travelled in a hybrid electric car or bus?

Engineers developed hybrid electric vehicles to save fuel and money, and to reduce greenhouse gases.

A **hybrid electric car** includes an internal combustion engine (like a normal car) and a big battery. At higher speeds, the engine propels the car. At lower speeds, the battery takes over.

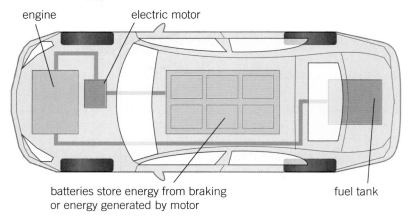

▲ A hybrid electric car viewed from above.

▲ A hybrid electric bus.

A State what a hybrid electric car is.

Pros and cons of different car types

In a petrol car, fuel burns in the engine. The engine turns a transmission. This turns the wheels. In an electric vehicle, a battery provides electricity to an electric motor. The motor turns the transmission. This turns the wheels.

The table shows some advantages and disadvantages of petrol and electric cars.

Petrol car	Electric car
Travels a long distance on one tank of fuel.	Travels only 150 miles between charges.
Quick to refill.	Slow to charge.
Makes pollution as it moves.	Does not make pollution as it moves.
	The electricity used to charge the battery may have been produced from fossil fuels. This produces pollution.
Noisy.	Quiet.

B Describe one advantage and one disadvantage of an electric car.

How do hybrid electric cars work?

Hybrid electric cars combine the best features of petrol and electric cars. There are two main types:

- In a parallel hybrid, a fuel tank supplies fuel to the engine. Batteries supply chemical energy to the electric motor. The engine and motor can turn the transmission at the same time.
- In a series hybrid, petrol provides energy to turn a generator. Sometimes, the generator charges the batteries. The rest of the time, the generator provides electricity to a motor. The motor turns the transmission.

C Name one component that is in a series hybrid car but not in a parallel hybrid car.

Why do hybrid cars use less fuel?

Hybrid electric cars use less fuel than petrol cars. This means they create less pollution. There are several reasons for this, including:

- A hybrid car does not need the engine to turn all the time because it has an electric motor. The car may automatically turn off the petrol engine at a red light.
- Hybrid cars use regenerative braking. Normally, when a car brakes, kinetic energy is dissipated and heats the surroundings. In a hybrid car, this energy recharges the battery.

What's in a battery?

Most hybrid car batteries are nickel–metal hydride batteries. Nickel may cause cancer. Scientists are working on making lithium-ion batteries for cars, like the batteries in your phone or laptop.

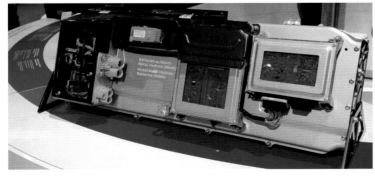

▲ A nickel–metal hydride car battery.

Comparing cars
Write the text and draw pictures for a webpage to help people decide whether to buy a hybrid electric car or an electric car.

Key Words
hybrid electric car

Summary Questions

1. Copy and complete the sentences below.
   A petrol car has an internal _____ engine. An electric car has a rechargeable _____. A _____ electric car has both of these. A hybrid car burns _____ fuel than a petrol car and produces _____ exhaust gas.
   (5 marks)

2. Describe two advantages and two disadvantages of petrol-driven cars and of electric cars.
   (4 marks)

3. Draw a visual summary that explains the differences between petrol-driven cars and hybrid cars.
   (6 marks)

C3 Chapter 1 Summary

Key Points

- Nanoparticles are tiny pieces of a substance with a diameter of between 1 nm and 100 nm.
- 1 g of a nanoparticle substance has more atoms on its surface than 1 g of the same substance existing in normal-sized pieces.
- Carbon nanotubes are cylinders of carbon atoms.
- Carbon nanotubes are strong, stiff, and have low densities. They make items strong, stiff, and lightweight.
- Nanoparticles can be used to protect items from water and stains. Nanomedicine is the use of nanoparticles to treat disease. Scientists are developing the use of nanoparticles to release antibiotics in bandages, to deliver drugs to tumours, and to destroy tumours.
- Scientists are investigating the risks of using nanoparticles.
- Most cars burn petrol or diesel. These fuels are mixtures of hydrocarbon compounds.
- Burning petrol or diesel produces two main products – carbon dioxide and water.
- Other substances are produced by chemical reactions in car engines, including nitrogen oxides, carbon monoxide, particulates, and unburned hydrocarbons.
- Chemical reactions in catalytic converters remove carbon monoxide, nitrogen monoxide, and unburned hydrocarbons from vehicle exhausts.
- Hydrogen can be used as a vehicle fuel. It produces one combustion product – water.
- Hydrogen is difficult to store and transport safely.
- Hybrid electric cars include an internal combustion engine, fuelled by petrol or diesel, and a battery.
- A hybrid electric car uses less fuel than a petrol or diesel car.

Key Words

nanoparticle, nanometre, carbon nanotube, nanomedicine, safety, hydrocarbon, particulate, hydrogen fuel cell, biofuel, renewable, carbon neutral, catalytic converter, hybrid electric car

Maths challenge

Best way to travel?
A family of four is travelling from Exeter to Newcastle – a distance of 420 km and back again. They can travel by air, coach, train, petrol car, or hybrid car. The aeroplane, coach, and train will make the journey even if the family decide on another method of travel.

Task
Write a letter to the family recommending which method of travel they should use from an environmental point of view.

Tips
- Use the data in the table to estimate the mass of carbon dioxide produced for each travel option.
- Calculate the total for the whole family.
- Remember to include the return journey.

Method of transport	Mass of carbon dioxide produced (g/km)
air	170 (per person)
rail	53 (per person)
coach	30 (per person)
petrol car	150 (per car)
hybrid car	49 (per car)

End-of-chapter questions

1. Copy the table. Then tick one or more boxes to show which statement or statements are true of nanoparticles.

Statement	Is the statement true of nanoparticles?
Their diameter is between 1 nm and 1000 nm.	
100 g of a nanoparticle substance has a greater surface area than 100 g of the same substance as normal-sized pieces.	
A nanoparticle is made up of a few hundred atoms.	

(3 marks)

2. Copy and complete the sentences below. Use the words and phrases in the box once, more than once, or not at all.

   > biofuel, carbon dioxide, maize, water, sugar cane, oxygen

   Ethanol is an example of a _____. It is produced from crops such as _____ and _____. As these crops grow, they take in _____ gas from the air. An ethanol molecule is made up of atoms of carbon, hydrogen, and oxygen. This means that ethanol produces two main products when it burns: _____ and _____.

   (6 marks)

3. Read the passage in the box, then answer the questions that follow.

   > Medical staff use cotton gauze to help stop wounds bleeding. Medical gauze did not change much during the last century. In the early 2000s scientists developed a better gauze. First they added a substance called QuickClot to medical gauze. It stopped the bleeding but it became hot when in contact with blood or water. This caused burns. The scientists tried adding a different material to medical gauze — kaolin clay. The clay is rich in aluminosilicate nanoparticles. These help blood to clot, and so stop bleeding. The nanoparticles do not enter the body, since they get trapped at the injury site.

   a. Describe and explain the hazard linked to adding QuickClot to medical gauze. *(2 marks)*
   b. Name the substance in kaolin clay that helps blood to clot. *(1 mark)*
   c. Suggest why scientists working on the project claim that adding kaolin clay to medical gauze is unlikely to be hazardous to health. *(1 mark)*

   (4 marks)

4. Hexadecane is one of the compounds in diesel.
   a. Hexadecane is a hydrocarbon. State what is meant by a hydrocarbon. *(1 mark)*
   b. Write the names of the two main products formed when hexadecane burns. *(2 marks)*
   c. Write a word equation for the combustion reaction of hexadecane. *(2 marks)*
   d. The formula of hexadecane is $C_{16}H_{34}$. Give the number of atoms of the two elements that make up a molecule of hexadecane. *(2 marks)*
   e. Burning 1 kg of hexadecane releases 47 MJ of energy. Calculate the energy released on burning 3 kg of hexadecane. *(2 marks)*

   (9 marks)

5. Explain in detail how catalytic converters make car exhaust fumes less harmful.

   (6 marks QWC)

2.1 Evidence for atoms

Learning objectives

After this lesson you will be able to:
- describe evidence for Dalton's atomic model
- explain one way of developing a scientific explanation.

▲ How many pieces could you cut this chocolate into?

▲ Dalton collected gas from marshes to use in his experiments.

Imagine a bar of chocolate. How many pieces can you divide it into? Can you go on cutting it up forever?

People wondered about questions like this more than 2500 years ago. Some argued that matter completely fills its space. You can divide it into smaller pieces for ever.

But Indian and Greek philosophers had a different idea. They said that everything is made up of tiny particles, with empty space between. You cannot see the particles, or cut them up. The Greeks called these particles **atoms**.

The philosophers did not do experiments. They worked out their answers using creative thought.

▲ The Greek philosophers Leucippus and Democritus said that atoms are the smallest pieces of matter that exist.

A How did Indian and Greek philosophers work out that everything is made from atoms?

How did Dalton develop his atomic model?

For many years, there was no experimental evidence for atoms. Then John Dalton came along. He found evidence for atoms. His work marks a turning point in chemistry.

Dalton taught chemistry in Manchester. He was also interested in the weather. In 1787 he made the first of 200 000 entries in his weather diary. Dalton thought carefully about his observations.

He read about the work of a French chemist, Antoine Lavoisier. Lavoisier had discovered that the air included at least two gases.

Dalton worked out that water exists in the air as a gas, separate from other gases. He asked a question:

How can water vapour and air occupy the same space at the same time, when two substances in the solid state cannot?

Dalton suggested an explanation to answer his question. He said that water vapour and air are made up of separate particles. When water evaporates, its particles mix with air particles.

Dalton wanted to test his explanation. He designed and carried out experiments on mixtures of gases. The evidence supported his explanation. Dalton made a conclusion. In any mixture of gases, the particles of the different gases are not joined together. They are just mixed up.

Atoms and evidence
Imagine that Dalton and Democritus could meet to discuss their ideas about atoms. Write the script for a conversation they might have. Then perform the conversation with a partner.

Link
You can learn more about solids, liquids, and gases in C1 1.2 States of matter

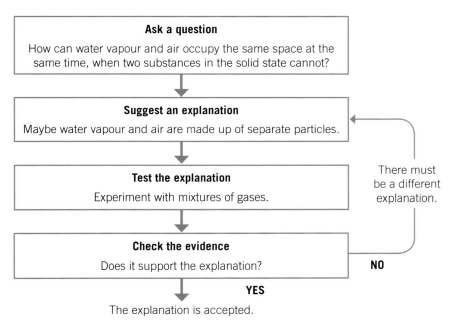

▲ Dalton found evidence for particles by asking a question and testing an explanation.

Dalton continued to experiment with gases. He extended his original explanation to develop an atomic model. His model includes these ideas:

- Each **element** has its own type of atom.
- The atoms of different elements have different masses.
- **Compounds** form when atoms of different elements join together.
- Atoms are re-arranged in chemical reactions.

B List the four ideas in Dalton's atomic model.

Key Words
atom, element, compound

Summary Questions

1. 🧪 Copy the sentences below, choosing the correct bold words.

   Everything is made up of **big/small** particles. There is **empty space/air** between the particles. Two gases can occupy the same space because their particles are **joined together/mixed up**.

   (3 marks)

2. 🧪🧪 Explain how Dalton used evidence from another scientist to help him develop his atomic model.

   (2 marks)

3. 🧪🧪🧪 Compare the ideas of Democritus and Dalton, and the ways in which they developed their ideas.

   (6 marks QWC)

2.2 Looking into atoms

Learning objectives

After this lesson you will be able to:
- explain how scientists discovered electrons and the nucleus.

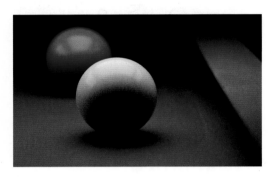

▲ Dalton imagined atoms as tiny billiard balls.

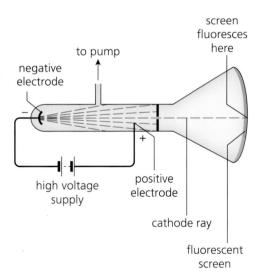

▲ Scientists used this apparatus to investigate the effect of electricity on gases.

Link

You can learn more about electric charge in P2 1.1 Charging up

Is there anything inside an atom? Is it a solid sphere? Or is it made up of even tinier particles?

Dalton imagined an atom as a billiard ball. You cannot break it up. It is the same all the way through. Other scientists questioned this idea. By 1904, they had some answers.

A State how Dalton imagined an atom.

What's in an atom?

Scientists were investigating gases. They put a tiny amount of gas in a sealed tube with a screen. They built an electric circuit, and supplied a big voltage. Part of the screen glowed green. Why? Scientists suggested that rays from the negative electrode caused the green glow. The rays travelled through the gas and hit the screen.

Scientist J. J. Thomson asked a question about the rays:

What are the rays made of?

He wondered if the rays were electrically charged. He decided to test his explanation.

Thomson passed the rays between two pieces of metal with electrical charges. The rays bent towards the positively charged metal. Thomson's explanation was correct. The rays are negatively charged.

Thomson did more experiments. He thought about the evidence. He realised that the negative rays are made up of tiny particles. He called the particles **electrons**. Every electron has the same negative charge and tiny mass.

Thomson wondered where the electrons came from. He realised that, since everything is made up of atoms, electrons must be part of atoms.

B State what an electron is.

In 1904 Thomson suggested a new atomic model. He said that an atom has negative electrons moving around in a positively charged sphere. Some people called Thomson's idea the plum-pudding model. It reminded them of plums in a pudding.

c Describe the plum-pudding model of an atom.

Is the plum-pudding model correct?

Ernest Rutherford wondered if the plum-pudding model was correct. He worked with other scientists to find out. The scientists set up the apparatus below.

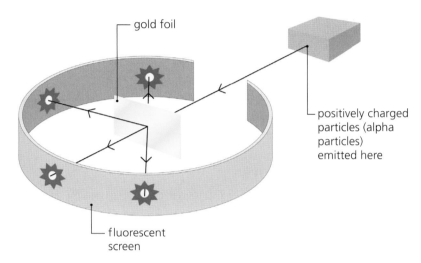

▲ Rutherford, Geiger, and Marsden used this apparatus to test the plum-pudding model.

The scientists planned to fire positive particles at the gold foil. They predicted that most would go straight through. A few would pass close to negative electrons and be attracted towards the electrons; they would change direction slightly.

The scientists carried out their plan. The results were bizarre. Some positive particles bounced back off the foil.

Rutherford thought about the strange results. How could he explain them? By 1911, he had a new model:

- Atoms have a central nucleus with a positive charge. Most of the mass of the atom is here.
- There is a big empty space around the nucleus. Electrons move around here.

How massive?

The nucleus of an oxygen atom has a relative mass of 16. An electron has a relative mass of 0.0005. How many times more massive is an oxygen atom nucleus than an electron?

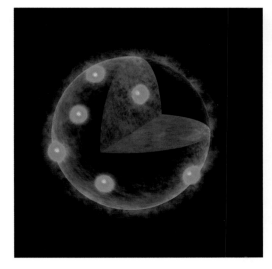

▲ Thomson's plum-pudding model.

Key Words

electron

Summary Questions

1. Copy and complete the sentences below.

   Dalton imagined an atom as a billiard _____. J. J. Thomson created the plum-_____ model of an atom. It had _____ electrons moving around a positively charged sphere. Rutherford discovered that an atom has a central _____ with a _____ charge.

   (5 marks)

2. Explain how Rutherford discovered that an atom has a positively charged nucleus.

   (4 marks)

3. Make a chart to summarise the models described on these two pages, and the evidence for them.

   (6 marks)

2.3 Discovering the Periodic Table

Learning objectives

After this lesson you will be able to:
- describe how Mendeleev devised the Periodic Table
- describe how Tacke and Noddack discovered rhenium.

By 1860 scientists knew of around 60 elements. They knew a great deal about their properties. But much of this knowledge was a jumble of facts.

Russian chemist Dmitri Mendeleev sparked another turning point in chemistry. He devised the **Periodic Table**. The Periodic Table is a vital tool in chemistry. It shows patterns in properties, and helps us to make predictions.

A Name the scientist who devised the Periodic Table.

Organising elements

In 1860 Mendeleev went to a conference in Germany. He heard a talk by an Italian, Stanislao Cannizzaro. Cannizzaro had worked out the masses of the atoms of the elements known at the time. He gave out copies of his data. Mendeleev picked up a copy.

Several chemists tried using atomic mass data to organise the elements, and show patterns in properties. No-one came up with the perfect arrangement.

One day Mendeleev was at home, writing a textbook. He was trying to organise information about elements' properties. Mendeleev cut out some cards. On each card he wrote the name of an element and some data for the element, including:

- its atomic mass
- its physical properties
- the formulae of its compounds with hydrogen and oxygen.

Mendeleev tried sorting the cards in different ways. Eventually, he came up with an arrangement that worked. The elements were in atomic mass order. Elements with similar properties were grouped together. Mendeleev wrote his arrangement on the back of an envelope. This was the first Periodic Table.

Mendeleev was confident in his arrangement. He left gaps for elements that he predicted should exist but that had not yet been discovered. He predicted the properties of these elements. Later, chemists discovered the missing elements, including gallium, scandium, and germanium.

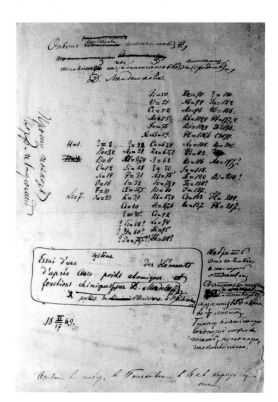

▲ Mendeleev's first Periodic Table.

Key Words

Periodic Table, catalyst

B State why Mendeleev left gaps in his Periodic Table.

Who discovered rhenium?

By 1925 most of the gaps in the Periodic Table had been filled. But there were spaces beneath manganese for two undiscovered elements.

Two German chemists, Ida Tacke and Walter Noddack, extracted 1 g of a new substance from 660 kg of an ore. The chemists thought the substance might be one of the missing elements.

Tacke and Noddack asked Otto Berg to do some tests on the substance. He confirmed that the substance was one of the missing elements. They called the element rhenium.

Rhenium speeds up chemical reactions – it is an excellent **catalyst**. It is also included in alloys for jet engine parts.

▲ Mendeleev predicted the existence of scandium, and its properties.

Fantastic Fact

Ida Tacke was nominated three times for the Nobel Prize in Chemistry but never won it.

◄ Ida Tacke was one of the discoverers of the element rhenium.

C Name the three scientists involved in the discovery of the element rhenium.

Mendeleev's musings

Imagine what Mendeleev might have said when he first devised the Periodic Table. Write a script, and perform it to a partner.

Summary Questions

1. Copy the sentences below, choosing the correct bold words.

    Mendeleev collected data about the **masses/sizes** of atoms of elements, their **physical/chemical** properties, and the formulae of their compounds with **nitrogen/oxygen** and hydrogen. He arranged the elements in order of increasing atomic **size/mass**. Elements with **similar/different** properties were grouped together.
    (5 marks)

2. Describe how Tacke and Noddack discovered a missing element, and why they asked another scientist for help.
    (3 marks)

3. Explain how Mendeleev devised the Periodic Table. Include an explanation of how he used data from another scientist.
    (6 marks QWC)

2.4 Lessons from fossils

Learning objectives

After this lesson you will be able to:
- describe how fossils are formed
- explain what fossils tell us about the ages of rocks.

What secrets do fossils hold?

◀ This is a 3.7-million-year-old fossil of human footprints at Laetoli, Tanzania. This shows that humans began walking much earlier than previously believed.

How do fossils form?

Fossils are the remains, or traces, of plants or animals that lived many years ago. They have been preserved by natural processes.

Most plants and animals do not form fossils when they die. An animal eats them, or they rot away. But occasionally a dead animal or plant (or its feces or footprints) is quickly covered in sand, mud, or dust – perhaps when a river floods or a volcano erupts. Then a fossil may form. The diagrams below show one way of forming fossils.

A Write down what a fossil is.

▲ Mary Anning investigated fossilised feces from ichthyosaurs. She broke them open and found fossilised fish bones. This shows what the dinosaurs ate.

▲ An animal dies. It falls onto mud or sand.

▲ More mud or sand quickly buries the body.

▲ Bacteria slowly break down the soft parts of the body. Its skeleton remains.

▲ The mud or sand above and around the skeleton starts to become rock.

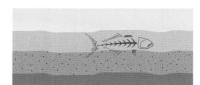

▲ As rock forms, underground water that is rich in dissolved minerals seeps into tiny spaces in the skeleton. These minerals gradually replace the original minerals of the skeleton. A hard copy of the original skeleton is formed.

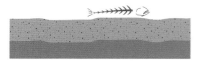

▲ Many years later, soft rock around the fossil is eroded. The fossil is exposed.

C3 Chapter 2: Turning points in chemistry

What do fossils tell us about their rocks?

Fossils are found in sedimentary rocks. Sedimentary rocks are formed in layers, over millions of years. The oldest layers are on the bottom. The layers are called **strata**.

Two hundred years ago a miner, William Smith, noticed that different strata contained fossils of different species. He asked a scientific question, and suggested an explanation:

Why do different rock strata contain different fossils? Perhaps rock strata of the same age always contain fossils of the same species…

Smith travelled widely looking for evidence. He published his findings so that other scientists could help collect evidence. Eventually he came up with an explanation:

Rock strata of the same age contain fossils of the same species. You can only find certain fossils in rock strata of a certain age.

Other scientists used Smith's explanation to classify rock strata by the fossils they contained. By 1850 they had divided the Earth's history into geological time periods. Each time period had its own **index fossil**. The index fossil shows the relative age of the rock.

Geological time period	Index fossil	Approximate date for start of time period (million years ago)
Quaternary		2.6
Tertiary		66
Cretaceous		146
Jurassic		200
Triassic		251

▲ Index fossils for the five most recent geological time periods.

B State what an index fossil shows.

Explaining explanations
Draw a flow diagram to show the stages William Smith went through to develop his explanation.

Link
You can learn more about rocks in C2 4.4 The rock cycle

Key Words
fossil, strata, index fossil

Foul Fact
Mary Anning was famous for her work on fossils. However, she was not allowed to join the Geological Society of London because she was female.

Summary Questions

1. 🧪 Copy and complete the sentences below.

   Fossils are the remains, or _____, of animals or _____ that lived many years ago. They have been preserved by _____ processes. Fossils form in _____ rock. Layers of rock are called _____.

   (5 marks)

2. 🧪🧪 Describe one way in which a fossil is formed.

   (4 marks)

3. 🧪🧪🧪 Make a poster to display some important lessons we have learned from fossils.

   (6 marks)

2.5 The oldest primate

Learning objectives

After this lesson you will be able to:
- explain how a recent fossil find makes scientists question earlier explanations about evolution
- describe the process of peer review.

In 2013 scientists announced an exciting find – a 55-million-year-old fossil.

It is the oldest **primate** fossil ever discovered. Primates are a group of mammals that includes monkeys, apes, tarsiers, and humans.

The team announced their discovery in a **scientific journal**. A scientific journal is a collection of articles written by scientists about their research. The articles are **peer reviewed**. This means that they are checked by other scientists who were not involved in the work but who are experts in that particular area of research. Scientists publish their work in journals to tell others about their research.

A State what a scientific journal is.

How did scientists study the fossil?

A Chinese farmer found a tiny fossilised skeleton in 2003. The farmer showed his find to a scientist, Dr Ni Xijun. A 10-year quest then began to solve the mysteries locked within the remains.

Dr Ni got together a team of experts from China and the USA. The scientists split apart the layers of rock that contained the fossil. They took accurate measurements and made detailed observations. Then they took the fossil to France. Another scientist used a new technique to study the fossil.

B State why the scientists took the fossil to France.

What can we learn from the fossil?
The life of *Archicebus*

Measurements show that the animal was just 71 mm long. Its mass was about 30 g. It would fit into the palm of your hand.

Scientists realised that the skeleton belongs to a species never seen before. They called the species *Archicebus achilles*. *Archicebus* had feet like a small monkey. Its skull had small eye sockets, suggesting it was active in daylight.

At the time *Archicebus* lived on Earth the planet was covered in tropical rainforests. *Archicebus*'s skeleton shows that it was active and agile. It leapt through the treetops. *Archicebus* had small, pointy teeth, perfect for eating insects.

▲ The 55-million-year-old fossil.

▲ Scientists think *Archicebus* looked like this.

C3 Chapter 2: Turning points in chemistry

C Explain why scientists think the animal was active in daylight.

Human evolution

At the moment, no-one knows whether the discovery will mark a turning point in the understanding of human evolution. But it seems likely that it will.

The scientists say that *Archicebus* is an ancestor of today's tarsiers. These are small, nocturnal primates. They live in Asian forests.

The scientists say that *Archicebus* changes earlier ideas that the first primates were the size of modern monkeys. Instead, they were probably small mammals, scurrying through rainforest canopies.

The scientists also point out that the new find may show that the first humans did not evolve in Africa – as scientists previously believed – but in Asia.

Finally, the scientists say that *Archicebus* shows that primates separated into two groups (tarsiers, and the group including monkeys, apes, and humans) up to 10 million years earlier than previously thought.

▲ A modern tarsier.

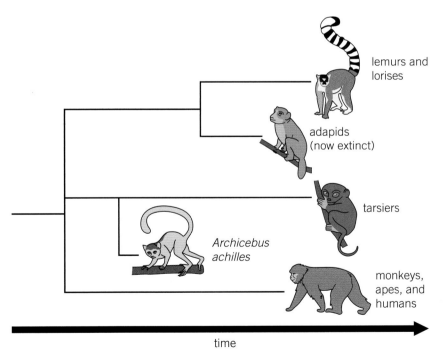

▲ *Archicebus* in the primate family tree.

Family tree

Look carefully at the family-tree diagram. With a partner, discuss what it shows. Then write an explanation of the diagram.

Key Words

primate, scientific journal, peer review

Summary Questions

1. Copy and complete the sentences below.
Scientists found the fossilised skeleton of a 55-million-year-old primate. Primates include monkeys, apes, tarsiers, and _____. The scientists published their findings in a scientific _____. Their article was peer _____.
(3 marks)

2. Explain why scientists think *Archicebus achilles* ate insects.
(2 marks)

3. Explain what scientists have learned from *Archicebus achilles*.
(6 marks QWC)

81

C3 Chapter 2 Summary

Key Points

- In the 1790s John Dalton found experimental evidence for the existence of atoms.
- John Dalton developed an atomic model. He imagined atoms as billiard balls.
- Dalton's model states that every element has its own type of atom, and the atoms of different elements have different masses. Compounds form when atoms of different elements join together. Atoms are re-arranged in chemical reactions.
- J. J. Thomson discovered that atoms are made up of even smaller particles, including electrons. An electron has a negative charge.
- Ernest Rutherford fired positive particles at gold foil. A few particles bounced back. Rutherford explained his observations by stating that most of the mass of an atom is in a central, positively charged nucleus.
- Stanislao Cannizzaro published a list of the masses of atoms of the 60 elements known in 1860.
- Dmitri Mendeleev used atomic masses and data on the properties of elements to arrange elements in the Periodic Table.
- The Periodic Table shows patterns in properties, and enables us to make predictions.
- Mendeleev left gaps for elements he expected would be discovered later. Ida Tacke and Walter Noddack discovered rhenium in 1925.
- Fossils are the remains, or traces, of plants and animals that lived many years ago. They have been preserved by natural processes.
- Fossils form in sedimentary rocks. The layers in sedimentary rocks are called strata. Rock strata of different ages contain fossils of different species.
- Scientists report their findings in scientific journals. Articles in scientific journals are peer reviewed.

Key Words

atom, element, compound, electron, Periodic Table, catalyst, fossil, strata, index fossil, primate, scientific journal, peer review

Case study

Exposing explanations
Scientists use evidence and their scientific knowledge to develop explanations.

Task
Write a plan for a lesson that will teach students in Year 7 how scientists develop scientific explanations.

Tips
- Create a table to show each of the stages below for every explanation that is described:
    - asking a question
    - suggesting an explanation
    - testing the explanation.

End-of-chapter questions

1. 🧪 Fossils provide clues about life on Earth in earlier times. Copy the descriptions of fossil discoveries. Then write the letter of one explanation next to each description.

   (4 marks)

Description of discovery	Explanation
bones from smaller ichthyosaurs in fossilised ichthyosaur feces	
spiral markings on fossilised ichthyosaur feces	
fossilised fish bones in fossilised ichthyosaur feces	
fossilised ichthyosaur feces that are black in colour	

   Explanations:

   A Ichthyosaurs ate sea animals.
   B Ichthyosaurs ate other ichthyosaurs.
   C Ichthyosaurs had ridges in their intestines.
   D Ichthyosaurs ate animals that contained ink sacs, like modern squid.

2. 🧪 Copy and complete the sentences below.

   Everything is made up of tiny particles called _____. Each of these particles has a _____ in its centre. This is _____ charged. There is an empty space around the centre of the atom. Tiny particles that are _____ charged move around here.

   (4 marks)

3. 🧪🧪 Copy the diagram below, which outlines how scientists may develop a scientific explanation.

   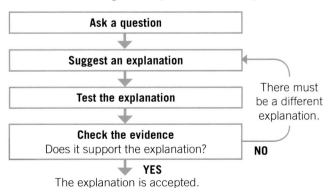

   Then write each of the three statements below in the correct box of the diagram to outline how Mendeleev discovered the Periodic Table.

   (3 marks)

   Statements

   X Write the elements in order of the mass of their atoms. Look for patterns in the elements' properties.

   Y What are the patterns in the properties of elements?

   Z When atoms are arranged in order of the mass of their atoms, there is a repeating pattern of properties.

4. 🧪🧪🧪 Read the passage in the box.

   > John Dalton knew that the air included at least two gases. If you had 1 m³ of each gas, they would each have a different mass. Dalton asked a question:
   >
   > *Why does gravity not separate the heavier gas from the lighter gas?*
   >
   > Dalton suggested a possible explanation to answer his question. The gas particles, he said, were surrounded by heat. Heat keeps the particles apart, and prevents them from settling down into separate groups. He extended his idea by assuming that each gas had its own type of atoms. The greater the mass of 1 m³ of a gas, the heavier its atoms.

   Write a critical review of Dalton's ideas described in the passage above. In your answer, state which parts of the explanation are correct, and improve any parts of the explanation that are not correct.

   (6 marks QWC)

3.1 Break-in!

Learning objectives

After this lesson you will be able to:
- explain how to separate mixtures by filtration and evaporation.

There has been a break-in at Sava Shop. The window is broken. Outside, there is blood on the sandy path. Inside is a horrible stink. Has someone been sick?

▲ The crime scene.

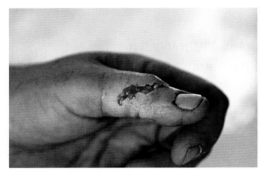

▲ The driver's hand.

Along the road, police stop a swerving car. The driver has blood on his hands and damp patches on his coat. His breath smells of cider. More noticeable is another, stronger smell. Could it be vomit, or rancid sweat?

A Name the sense the police officers use to detect cider.

What evidence do the police collect?
From the driver

At the car, police officers breathalyse the driver. The test is positive. He has drunk too much alcohol to legally drive. The officers take the driver to the police station. Here, he gives his name – Ryan. Ryan takes off his coat. An officer places the coat in a bag, and seals it up. The foul smell is less strong.

The officer notices cuts on Ryan's hands. She uses a damp cotton-wool swab to collect some dried blood. She places the swab in pure water, seals the container, and puts it in a freezer. Once frozen, she will send the sample for **DNA** analysis. Does the blood belong to Ryan or to someone else?

Later, a doctor comes to the police station. She takes 10 cm^3 of blood from a vein in Ryan's arm, and separates it into two parts. This blood will be tested for alcohol.

Key Words

DNA, filtration, filtrate, evaporate

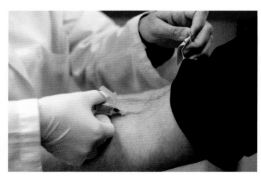

▲ Taking the blood sample.

C3 Chapter 3: Detection

From the shop
Scene-of-crime officers arrive at the shop. They take samples of bloody sand from the path. They look for fingerprints and hairs.

The smell inside the shop is overpowering. The police call the fire brigade. Is it safe to enter? Can they find out what is causing the smell?

How can you separate blood from sand?
The crime-scene investigators send the bloody sand to a laboratory. They hope to use the blood to help identify the criminal.

Scientists use **filtration** to separate the sand and blood. They add pure water to the bloody sand. The blood mixes with the water. They filter the mixture.

Link
You can learn more about separation techniques in C2 2.4 Filtration

Media briefing
You are a senior police officer. You need to brief the media about the crime. Plan what to say, then deliver your briefing. Speak calmly and clearly.

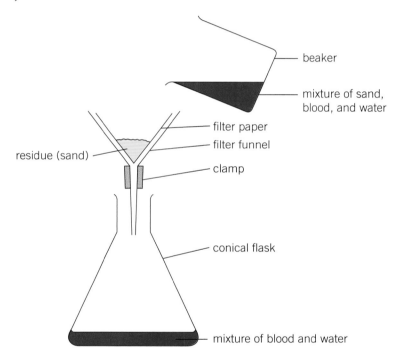
▲ Filtering the mixture.

When the mixture is filtered, sand remains in the filter paper. The grains of sand are too big to get through the tiny holes in the filter paper. The particles in the water and in the blood mixture are much smaller than the grains of sand. They go through the tiny holes in the filter paper, into the flask beneath.

The scientists keep the **filtrate**. They allow the water to **evaporate**. They plan to do a DNA test on the blood. Will the sample be good enough to help identify the criminal?

B Name the two processes used to separate sand from bloody water.

Summary Questions

1. Copy the sentences below, choosing the correct bold words.

   You can use **filtration/evaporation** to separate sand from a solution. You can then use **filtration/evaporation** to remove water from the solution.

   (2 marks)

2. Explain how filtration works.

   (3 marks)

3. Suggest why the smell from Ryan's coat is less strong when sealed in a bag.

   (2 marks)

4. Draw and annotate a visual summary of the evidence collected so far. In a different colour, write down what the police might find out from each piece of evidence.

   (6 marks)

3.2 Smelly problem

Learning objectives

After this lesson you will be able to:
- explain how to identify acids and alkalis
- give examples of neutralisation reactions.

Specially trained fire-fighters arrive at the shop. The smell is less strong, since some particles of the smelly substance have diffused through the broken window. Two fire-fighters put on gas-tight suits, gloves, and breathing apparatus. They enter the shop.

A State why the smell has become less strong.

Is the smelly liquid acidic?

Behind the window is a puddle of liquid. It looks oily, unlike water. Is this the source of the smell?

The fire-fighters dip red **litmus** paper into the liquid. There is no change. They do the same with blue litmus paper. The paper turns red. This tells them that the liquid is acidic.

They check the liquid with **universal indicator** paper. Its colour shows that the liquid pH is 3.

▲ This fire-fighter is wearing breathing apparatus.

▲ The liquid is acidic.

B State how the fire-fighters know the liquid is acidic.

Crime report

Write a newspaper report on the crime so far. Plan your work carefully. Decide what to include in each paragraph, and choose the best order for your paragraphs.

Key Words

litmus, universal indicator, hazard

What is the liquid?

The fire-fighters collect a sample of the liquid. They will use instruments in their vehicle to identify it.

At that moment, a police inspector arrives. He thinks he recognises the smell from a previous incident. Is it butanoic acid? A colleague looks it up on the Internet. She reports that butanoic acid has a foul smell – like vomit or rancid butter. Its melting point is −8 °C and its boiling point is 164 °C, so it is liquid at room temperature (22 °C). Its formula is $C_4H_8O_2$.

Butanoic acid has been used in acid attacks, says the colleague. A man threw the substance into a victim's face, blinding him. Protestors have used butanoic acid to attack whaling ships.

The fire-fighters get their results. The substance is butanoic acid.

▲ Protesters have used butanoic acid to attack whaling ships.

C Write down the state of butanoic acid at room temperature.

How can they dispose of the liquid?

Butanoic acid has this **hazard** symbol. It is corrosive.

Safety guidance for butanoic acid includes:

- Avoid contact with skin and eyes.
- Wear eye protection and gloves.
- Do not inhale the substance.
- Do not empty it into drains.

The fire-fighters look up how to deal with spilt butanoic acid. There are two options:

- For large amounts, add sand or soil to absorb the liquid. Then call in a specialist waste-disposal company to remove the mixture.
- For small amounts, add sodium hydroxide solution to neutralise the acid:

  butanoic acid + sodium hydroxide → sodium butanoate + water

  The salt produced in the reaction, sodium butanoate, is much less harmful than butanoic acid.

The puddle of liquid is quite big so the fire-fighters decide to cover it with sand.

D Name the products of the neutralisation reaction between butanoic acid and sodium hydroxide.

Summary Questions

1. Copy and complete the sentences below.
   If you dip blue litmus paper into an acid, the paper changes colour to _____. You can use universal indicator to find out the _____ of the acid. When an acid reacts with an alkaline solution, the products are a _____ and _____.
   (4 marks)

2. Copy and complete the word equations below.
   a. ethanoic acid + _____ hydroxide → potassium ethanoate + _____ *(2 marks)*
   b. propanoic acid + sodium hydroxide → _____ propanoate + _____ *(2 marks)*
   c. butanoic acid + potassium hydroxide → _____ _____ + water *(1 mark)*

3. Explain how neutralisation reactions produce substances that are safer than the starting materials. Illustrate your answer with examples.
   (6 marks QWC)

3.3 Message in a bottle

Learning objectives

After this lesson you will be able to:
- describe how chromatography separates dyes in ink.

The DNA analysis of the blood from Ryan's hand has been returned to the police station. It turns out that the DNA is not new to the police. It matches a sample analysed 10 years ago, found on a blood-stained note posted to the police. The note was written in fountain pen.

The police never found the writer, or the garden. What secrets might it hold? The police get permission to search Ryan's house. Is something buried in the garden? Will they find the ink used to write the note?

▲ DNA test results.

How did they analyse the ink?

At Ryan's house, the police do a careful search. At the back of a kitchen cupboard, they find an old bottle of ink.

The police officers know they must send the ink to the forensic laboratory. But they do their own test first, using a technique called **chromatography**.

▲ The ink.

The officers take a piece of chromatography paper. They draw a line on the paper, in pencil. They drip a drop of the ink onto the line. They place the bottom of the paper in a beaker of water. The officers wait. As soon as the first spot reaches the top of the paper, they remove the chromatogram.

A Name the technique used to analyse the ink.

How chromatography works

Water moves up the paper. It reaches the ink. Ink is a mixture of dyes. Some dyes mix well with water. Others mix less well. Some dyes are attracted to the paper more strongly than others.

The dyes travel up the paper at different speeds. The dye that travels fastest reaches the top of the paper. This dye mixes well with water, and is not strongly attracted to the paper.

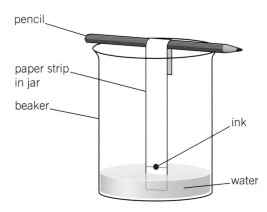

▲ The officers' chromatography apparatus.

C3 Chapter 3: Detection

In paper chromatography, the paper is the **stationary phase**. The water is the **mobile phase**.

B Name the mobile phase and the stationary phase in paper chromatography.

Key Words
chromatography, stationary phase, mobile phase

What the chromatogram shows
The officers look back at the records from 10 years ago. They find an old chromatogram, made from ink extracted from the note. They compare this chromatogram with their own.

recent chromatogram from ink   chromatogram from old note

▲ The two chromatograms.

The police officers have different opinions about what conclusion they can make from the chromatograms.

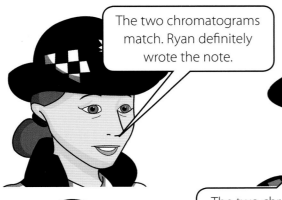

The two chromatograms match. Ryan definitely wrote the note.

The two chromatograms match. But lots of people use that brand of ink. Ryan might have written the note but we can't be sure.

The two chromatograms are not identical – the spots are different shapes. Ryan did not write the note.

Repeatable results?

The police officers want to check that their results are repeatable. They make some more chromatograms with the ink. Is this is a good idea? Can it help them decide whose conclusion is correct? Explain your answers.

Summary Questions

1. Copy and complete the sentences below, choosing the correct bold words.
   Paper chromatography separates substances in **compounds/mixtures**. The **paper/water** is the stationary phase and the **paper/water** is the mobile phase. The dye that mixes **best/least well** with the water and that is attracted **best/least well** to the paper travels fastest up the paper.
   (5 marks)

2. Read the police officers' different conclusions. Write down which conclusion you think is best, and give a reason for your choice.
   (3 marks)

3. Draw and annotate a diagram to explain how paper chromatography works.
   (6 marks)

3.4 Blood alcohol

Learning objectives

After this lesson you will be able to:
- describe how gas chromatography separates alcohol from blood.

▲ Injecting a sample for gas chromatography.

Legal limits

The table shows the maximum legal blood–alcohol levels for driving in different countries (June 2013). Plot the data on a bar chart, and write a paragraph to compare the values.

Country	Maximum legal blood–alcohol concentration for drivers (mg of alcohol in 100 cm^3 of blood)
Australia	50
Denmark	50
France	50
Hungary	0
India	30
Netherlands	50 (new drivers 20)
Poland	20
Sweden	20
UK	80
US	80

On the day the crime was discovered, a doctor used a syringe to remove 10 cm^3 of blood from Ryan's arm. She divided the blood into two, and sealed and labelled each sample. She gave one sample to a police officer, and one to Ryan. The police officer sent his sample to a laboratory. Ryan gave his sample to a solicitor, who put it in the fridge.

How do scientists analyse blood?

At the laboratory, a chemist looks carefully at the labels and seals on the container. She must be sure they have not been tampered with. She injects some blood into a **gas chromatography** instrument. Gas chromatography separates and identifies substances in mixtures and measures their amounts.

In the gas chromatography instrument, the blood sample is heated. Any alcohol in the blood becomes a gas. Other substances in the sample also turn to gas.

Helium gas is added to the mixture of gases. Helium is the mobile phase, like water in paper chromatography. The mixture flows through a tube. In the tube is a polymer column. The polymer is the stationary phase.

Different substances move through the column at different speeds. The speed at which a substance travels depends on how strongly it is attracted to the polymer.

A Name the stationary phase and the mobile phase in gas chromatography.

If a driver has had any alcohol at all there is an increased chance of having an accident, even if they are not over the UK blood–alcohol limit.

How do scientists interpret the chromatogram?

Coloured spots show separate substances in paper chromatography. Gas chromatography produces a different record, called a gas chromatogram. This has a peak for each substance the instrument detects.

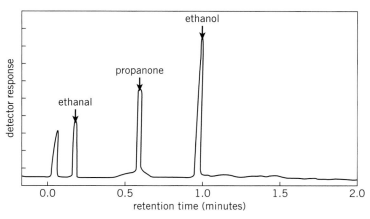

▲ Gas chromatogram of Ryan's blood.

B State what a gas chromatogram is.

The chromatogram shows that there is ethanol in Ryan's blood. Ethanol is the drug in alcohol. The chemist looks at the height of the ethanol peak and at the vertical scale. There is a high concentration of ethanol in Ryan's blood. It is higher than the concentration allowed for driving legally.

How did Ryan respond?

The chemist sends the test result to the police station. Ryan is not happy. "The result is wrong," he says. "My sample was muddled with someone else's."

Ryan sends his sample to a different laboratory. A chemist obtains the chromatogram below.

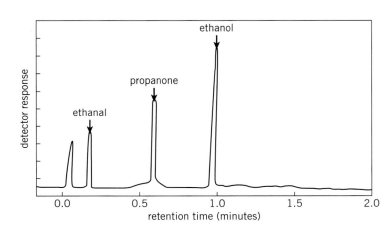

Key Words

gas chromatography

Summary Questions

1. Copy and complete the sentences below.

   A technique called _____ chromatography is used to identify alcohol in blood. In this type of chromatography, helium gas is the _____ phase, and a _____ column is the stationary phase. A gas chromatogram shows a _____ for each substance the instrument detects.

   (4 marks)

2. Describe how different substances are separated from blood in gas chromatography.

   (4 marks)

3. Compare the chromatograms from the two laboratories. What conclusion can you make? Explain your answer.

   (4 marks)

4. Write a paragraph to compare paper chromatography and gas chromatography.

   (6 marks, QWC)

3.5 Body!

Learning objectives

After this lesson you will be able to:
- use the reactivity series to predict whether metals will react with oxygen and water vapour in the air or soil.

▲ Police officers dig up Ryan's garden after discovering the note.

metal, reactivity series, rust

The police officers look back at the note. "Buried in my garden," it says. The police officers find out Ryan's addresses of the past 15 years. They get permission to dig up all three gardens. What will they find?

What's under the flowers?

The police officers go to Ryan's current address. They dig up the garden. They find nothing odd.

Next, the officers go to a previous address. They dig carefully. Under a flowerbed, they find a small bone. Is it from a dead cat or dog? Or might it be something more sinister?

The officers continue digging. They find more bones, wrapped in what might once have been carpet. They look more closely. The bones are human.

The flesh has decayed. No-one could recognise the person. But around one finger is a shiny gold-coloured ring. Around the neck is a blackened necklace. Around a wrist is a rusty watch. Can these **metal** clues identify the body?

What does Ryan say?

The police officers question Ryan about their new finds. Does he know anything about the skeleton in his old garden?

Ryan keeps quiet, until the officers show him photos of the jewellery. "That proves it wasn't me," he says. "I moved away from that house 10 years ago. There is no way the ring could still be that shiny!"

▲ The rusty watch.

▲ The ring.

What do the metals tell us?

Gold is near the bottom of the **reactivity series**. It does not react with oxygen or water. A gold object remains shiny for many years, even when buried in damp soil.

potassium
sodium
magnesium
zinc
iron
lead
copper
silver
gold
platinum

▲ The reactivity series.

A Name one metal that does not react with oxygen or water.

The police officers show the other metal objects to a forensic chemist. The necklace is silver. Silver is near the bottom of the reactivity series but it does react with sulfur compounds in the air. The product is black silver sulfide.

The watch, says the forensic chemist, is mainly iron. It has rusted. **Rust** forms when iron reacts with oxygen and water. Rust is hydrated iron oxide, which is an orange-brown colour.

iron + water + oxygen → hydrated iron oxide

The officers consider the evidence. The metals found with the body could have been buried 10 years ago...exactly when Ryan lived at the house where the body was found. Has the shop break-in led them to find a murderer?

B Use the reactivity series to predict two metals that react easily with oxygen and water.

Link

You can learn more about the reactivity series in C2 3.4 Metal displacement reactions

Summary Questions

1 Copy and complete the sentences below, choosing the correct bold words.

The reactivity series shows how vigorously metals react. Gold is near the **top/bottom** of the reactivity series. It **reacts vigorously/does not react** with oxygen and water. Iron is **higher/lower** in the reactivity series. It reacts with oxygen and **water/nitrogen** to make rust. The scientific name for rust is **hydroxide/hydrated** iron oxide.
(5 marks)

2
a Name the most reactive metal in the reactivity series on this page. *(1 mark)*
b Use the reactivity series to name two metals that do not react with water or oxygen. *(1 mark)*
c Name the product formed when zinc reacts with oxygen. *(1 mark)*

3 Ryan will soon have to face a courtroom trial. Write a paragraph for the jury about the metal objects found with the body. Explain in detail how their condition suggests that they were buried 10 years ago.
(6 marks)

3.6 Clues in the carpet

Learning objectives

After this lesson you will be able to:
- describe the difference between biodegradable and non-biodegradable materials
- explain why the properties of sisal, wool, and poly(propene) make them suitable for carpets.

The crime-scene investigators investigate the carpet that was wrapped around the bones found in Ryan's garden. Could it have been buried 10 years ago?

What's in a carpet?

Different materials are used to make carpets. Some carpets have a woven sisal base. Wool tufts are joined to this base.

- Sisal is a **natural** product, obtained from a plant. Each leaf contains many fibres. They are extracted and made into strong threads. Sisal is an insulator. It does not catch fire easily.
- Wool is also a natural product, obtained from sheep. Like sisal, it is an insulator. It feels soft. It does not catch fire easily.

▲ Was this carpet buried 10 years ago?

▲ Wool is a natural product.

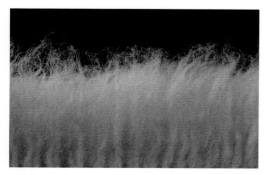
▲ A warm-feeling and soft wool carpet.

▲ A sisal plant.

A Describe two properties that make wool suitable for making carpets.

Some carpets include poly(propene) tufts. Poly(propene) is a **synthetic** polymer. It is made from substances obtained from crude oil. It is hard-wearing and easy to clean.

C3 Chapter 3: Detection

▲ Poly(propene) carpets can be dyed with many colours.

B Describe two properties that make poly(propene) suitable for making carpets.

Which carpet materials are biodegradable?

Some materials are **biodegradable**. This means that they can be broken down by natural processes. In the soil, bacteria or fungi will help them to rot away. Wool and sisal are biodegradable.

Poly(propene) is not biodegradable. It does not rot away naturally. It may remain unchanged in the environment for many years.

If the body was wrapped in a poly(propene) carpet, the carpet would not have changed much over time. Wool carpet, however, would have partly rotted away.

C State what the word biodegradable means.

What does the carpet tell us about the crime?

Forensic scientists look carefully at the carpet around the body. It has partly rotted away. Its base appears to be made of woven sisal, and tufts of wool. The scientists conclude that their observations agree with the idea that the carpet was buried 10 years ago, at the time the police received the note.

The evidence is fitting together...

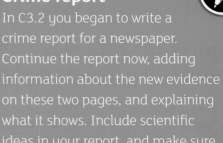

Crime report

In C3.2 you began to write a crime report for a newspaper. Continue the report now, adding information about the new evidence on these two pages, and explaining what it shows. Include scientific ideas in your report, and make sure you organise your report logically.

Key Words

natural, synthetic, biodegradable

Summary Questions

1. Copy and complete the sentences below.
   Sisal is a suitable material for making carpets because it is a good _____. It does not catch _____ easily. Wool also has _____ that make it suitable for carpets. It feels _____.
   Poly(propene) is a synthetic _____ that also has properties that make it suitable for carpets.
   (5 marks)

2. Describe the difference between biodegradable and non-biodegradable materials and give examples of each.
   (4 marks)

3. Suggest some advantages and disadvantages of making carpet from a non-biodegradable material.
   (6 marks QWC)

3.7 A week in court

Learning objectives

After this lesson you will be able to:
- describe how scientific evidence can help solve crimes.

The police officers look back at their missing-person records. Two people went missing 10 years ago, at the time the police received the note and the body was buried. Their names were Angela Scott and Shannon Smith. Might the body be one of theirs?

The officers call in Angela's relatives. They look at the ring. It had belonged to Angela. A dentist studies Angela's **dental records**, and examines the teeth in the buried skull. The records match the teeth. The body is Angela's.

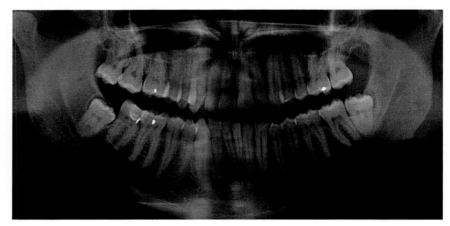

▲ Dental X-ray.

▲ The detectives gather all of the evidence and prepare their court case.

A State two methods the police used to identify Angela's body.

Preparing for court

The detectives have a great deal of evidence linking Ryan to the body:

- The DNA from the blood on the note matches the DNA in Ryan's blood sample and from his hands.
- The ink on the note matches the ink in Ryan's cupboard.
- The body was found in the garden of the house that Ryan was living in 10 years ago.

But is Ryan guilty of murder? Only a court can decide.

The detectives prepare their case carefully. Weeks later, the case comes to court. Ryan is charged with criminal damage to the shop, drink-driving, and the murder of Angela Scott.

B Name the three offences that Ryan has been charged with.

Key Words

dental record

Science solves crimes

Design a poster to show how the police officers, forensic scientists, and crime-scene investigators worked scientifically to solve the crime.

C3 Chapter 3: Detection

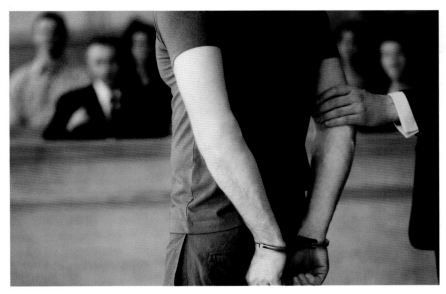

▲ The court room.

▲ Lawyers and police officers present evidence to the courtroom.

▲ The jury considers all the evidence they have heard. They use this to decide whether Ryan is innocent or guilty of the charges.

In court

The police officers give evidence in court. So does Ryan. The lawyers ask questions. They show that Ryan does not always tell the truth.

The officers tell the court that Angela and Ryan had been school friends. They saw each other regularly as friends after leaving school. 10 years ago, when they were both 19, they started dating. After a while, Angela decided to end the relationship. Ryan was very, very angry – angry enough to kill Angela.

Outside the courtroom, having heard all the evidence, the jurors discuss the evidence. They decide that the evidence is enough to convict Ryan of all three crimes. He is guilty, they say, beyond all reasonable doubt.

Summary Questions

1. Copy and complete the sentences below.

   There was a lot of _____ linking Ryan to the crime. The _____ from the blood on the note matched the DNA in Ryan's blood sample. The ink on the _____ matched the ink in Ryan's cupboard. The _____ was found in the garden of the house that Ryan was living in 10 years ago.

   (4 marks)

2. Describe the scientific evidence on this page that links Ryan to the murder of Angela Scott.

   (3 marks)

3. Describe two techniques used by detectives to obtain evidence used in Ryan's conviction.

   (6 marks)

4. Suggest why a jury might consider evidence from the buried skull and dental records to be more reliable than evidence from the ring buried with the body.

   (6 marks QWC)

C3 Chapter 3 Summary

Key Points

- Filtration separates an insoluble solid from a liquid, or from a solution of another substance.
- In filtration, pieces of solid do not go through the holes in filter paper. Particles of the liquid are small enough to pass through the holes.
- Red litmus paper becomes blue in an alkaline solution.
- Blue litmus paper becomes red in an acidic solution.
- The colour of universal indicator gives the pH of a solution.
- When an acid is neutralised by an alkali, the products are a salt and water.
- Paper chromatography separates the dyes in ink. The stationary phase is paper. The mobile phase is water or another solvent.
- In gas chromatography, the stationary phase is a polymer. The mobile phase is helium.
- A gas chromatogram has a peak for each substance the instrument detects. It shows the relative amounts of the substances in a mixture.
- The reactivity series lists metals in order of the vigour of their reactions.
- Potassium, sodium, and magnesium are near the top of the reactivity series. Copper, silver, and gold are at the bottom.
- Iron reacts slowly with oxygen and water to form hydrated iron oxide. This is rust.
- A biodegradable material is made harmless by natural processes. Bacteria or fungi may help the material to rot away. Biodegradable materials include wool and sisal.
- Poly(propene) is made from substances separated from crude oil. It is not biodegradable.

Key Words

DNA, filtration, filtrate, evaporate, litmus, universal indicator, hazard, chromatography, stationary phase, mobile phase, gas chromatography, gas chromatogram, biodegradable, metal, reactivity series, rust, dental record

Case Study

Biodegradable or not?
Some materials are biodegradable and some are not. How can you find out if a material will biodegrade?

Task
Plan an investigation to find out which of these materials are biodegradable and which are not: cotton, poly(ethene), polystryrene, paper, silk, wood.

Tips
In your plan:
- list the independent, dependent, and control variables
- draw a labelled diagram to show what to do
- make a results table to fill in
- state how you will know which materials are biodegradable, and which are not.

End-of-chapter questions

1. Write down the letter of the mixtures below that you could separate by filtration.
   a. sand from a mixture of sand and water
   b. copper sulfate from a solution of copper sulfate in water
   c. sand from a mixture of sand and salt solution
   d. salt from a mixture of sand and salt solution
   e. iron filings from a mixture of iron filings and water
   (5 marks)

2. A biodegradable material is one that it decays naturally. Three of the materials in the list below are biodegradable. Write down their names.

   wool   gold   flour   paper,   poly(propene)
   (3 marks)

3. The paper chromatogram below was obtained from a felt-tip pen.
   a. Name the stationary phase in paper chromatography. **(1 mark)**
   b. Look at the paper chromatogram below. Give the number of substances that were mixed to make the ink in the felt-tip pen. Explain your answer. **(2 marks)**

   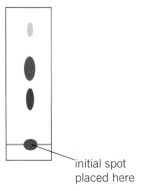

   (3 marks)

4. The gas chromatogram below was obtained from a mixture of substances.

   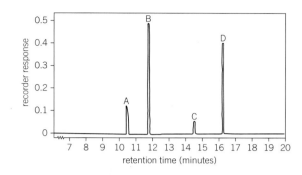

   a. Give the number of substances shown on the chromatogram. **(1 mark)**
   b. Give the letter of the substance present in the mixture in the greatest amount. **(1 mark)**
   c. Name the mobile phase in gas chromatography. **(1 mark)**
   (3 marks)

5.
   a. Name the three least reactive metals in the reactivity series. **(3 marks)**
   b. Name the most reactive metal in the reactivity series. **(1 mark)**
   c. Write a word summary for the reaction of iron with oxygen and water. **(2 marks)**
   (6 marks)

6. The equations below show some reactions of metals. Rewrite the equations so that they are balanced.
   a. $Cu + O_2 \rightarrow CuO$ **(2 marks)**
   b. $Ag + H_2S \rightarrow Ag_2S + H_2$ **(2 marks)**
   c. $Fe + O_2 \rightarrow Fe_2O_3$ **(2 marks)**
   d. $Na + H_2O \rightarrow NaOH + H_2$ **(2 marks)**
   (6 marks)

7. Compare the advantages and disadvantages of the techniques of filtration and gas chromatography to separate mixtures.
   (6 marks QWC)

Physics 3

In this unit you will learn about how technology, from mobile phones to hospitals, has changed the way that we live our lives. You will learn how our ideas about the Universe have changed, and how people discovered electromagnetism and radioactivity. You will also learn about how scientists look for aliens, new particles, and how GPS works.

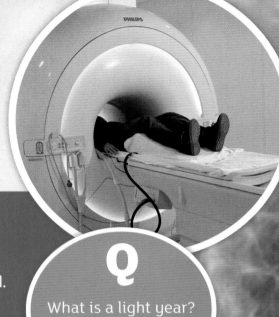

Q What is a light year?

You already know

- Resistance = potential difference//current
- A microphone converts a sound wave to an electrical signal.
- Light waves and infrared radiation are waves.
- Waves can be reflected and refracted.
- Speed = distance/time.
- The Moon is in orbit around the Earth, which is in orbit around the Sun, which is one of billions of stars in the Universe.
- The Earth is one planet in the Solar System, which is part of the Milky Way galaxy.
- You can make an electromagnet from a coil of wire, battery, and nail.
- Electricity is generated in a power station.

BIG Questions

- How does your mobile phone work?
- How did the Universe begin?
- How would we know if there were aliens out there?

Picture Puzzler
Key Words

Can you solve this Picture Puzzler?
The first letter of each of these images spells out a science word that you will come across in this unit.

Picture Puzzler
Close Up

Can you tell what this zoomed-in picture is?
Clue: It glows in the dark.

Making connections

In **P3** you will learn how the Universe began.

In **B3** you will learn about advances in medicine.

In **C3** you will learn more about fossils and our human ancestors.

1.1 Your phone

Learning objectives

After this topic you will be able to:
- describe the difference between an analogue and a digital signal
- describe how waves can be used for communication.

These days we rely on smart phones and other devices that communicate. How do they work?

Your call

You speak into your phone, and a microphone detects what you say. It converts the sound into an electrical signal.

The electrical signal is an **analogue** signal. It can have any value. The wave on the screen on the left is an analogue signal.

▲ You don't need a keyboard to operate this wearable computer.

▲ A microphone produces an analogue signal.

Link

You can learn more about how a microphone works in P1 2.4 Detecting sound

The digital age

Your phone converts the analogue signal from the microphone into a **digital** signal. It takes the amplitude of the sound wave at different times and converts it to a **binary** number. A binary number uses only 1s and 0s. In binary the number 3 would be 11. You can see other binary numbers in the table opposite.

A signal made of 1s and 0s is a digital signal.

Decimal number	Binary number (bit)
0	0000
1	0001
2	0010
3	0011
4	0100
5	0101
6	0110
7	0111
8	1000
9	1001
10	1010

A State the number of values that a digital signal can have.

Your friend's phone receives the digital signal and converts it back to an analogue wave. This is sent to a loudspeaker in the phone. She hears your voice.

Sending sounds as digital signals is a lot clearer than sending them as analogue signals.

Storing information

A 1 or a 0 is called a **bit** (short for **bi**nary digi**t**). If you put 8 bits together you get a byte. 1000 bytes = 1 kilobyte or 1 kB. 1 million bytes = 1 megabyte (1 MB), and 1 thousand million bytes = 1 gigabyte (1 GB).

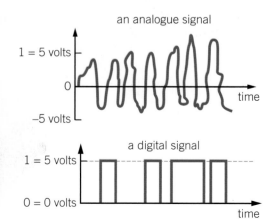

▲ An analogue signal can have any value, but a digital signal can have only two values.

When you store a song on your phone or laptop you store bits. Bits are stored like little pieces of magnetic material. If it is magnetised one way (say, N–S) it represents a 0. If it is magnetised the other way (S–N) it represents a 1.

B State the number of bits in a byte.

Sending and receiving

When you make a call or download a song you are transferring bits. That is why download speeds are calculated in bits per second. Information on CDs and DVDs is stored as a series of pits, which represent the bits.

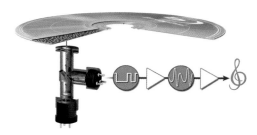

▲ A CD stores bits.

Using waves

Most smart devices don't need wires to communicate. The bits are sent or received using waves with the 1s and 0s coded into them. The waves are waves of the **electromagnetic spectrum**.

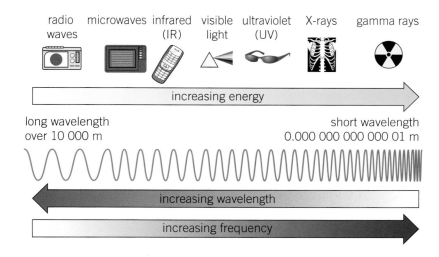

Devices that communicate via Bluetooth and Wi-Fi use **microwaves**. Infrared communication uses infrared waves.

All electromagnetic waves travel at the same speed as light and can travel through a vacuum. **Radio waves** have the longest wavelength and the lowest frequency. They are used to send radio and television signals.

Fantastic Fact

There is more computing power in a birthday card that sings 'Happy Birthday' than there was in the entire world 50 years ago.

Key Words

analogue, digital, binary, bit, electromagnetic spectrum, microwave, radio wave

Remember those waves!

Make up a mnemonic so that you can remember the waves of the electromagnetic spectrum.

Summary Questions

1. Copy and complete the sentences below.
   A smart phone uses a microphone to change sound to a signal that is _____. This is converted to a _____ signal made up of 1s and 0s. An _____ wave transfers information to another phone. Waves such as radio, microwaves and _____ are used for communication.
   (4 marks)

2. Calculate how many songs of size 1 MB you can fit on a 500 GB hard drive.
   (2 marks)

3. Compare analogue and digital signals and their uses.
   (6 marks QWC)

1.2 Your house

Learning objectives

After this topic you will be able to:
- describe what is meant by efficiency
- describe how an LDR detects light.

▲ Lightbulbs have changed over the years.

Link

You can learn more about how power stations generate electricity in P2 2.6 Energy resources

Think what houses were like 150 years ago, before electricity was invented. Can you imagine not being able to just switch a light on? What do you think the lighting will be like in houses of the future?

Light it up

Old types of bulb

Fifty years ago everyone used incandescent lightbulbs. These contained a filament that got so hot that it gave out visible light. One big problem was that most of the energy transferred was dissipated. The lightbulbs were not very **efficient**. Lots of energy was being used to heat up the room, not light it.

If a device is efficient most of the energy transferred to it does what we want. We want the lightbulb to light up the room, not heat it.

You can work out efficiency as a percentage. Devices such as lightbulbs, hair straighteners, or televisions cannot be 100% efficient. Some energy always ends up heating the environment. The incandescent lightbulb was only 10% efficient. 90% of the energy heated up the air.

$$\text{efficiency (\%)} = \frac{\text{useful energy (or power) out} \times 100}{\text{total energy (or power) in}}$$

A State the efficiency of an incandescent lightbulb.

Better bulbs

Scientists worked to make lightbulbs more efficient. They invented a **light-emitting diode** or **LED**. LEDs are about 30% efficient. They are very versatile and can be fitted into clothing or phones, and made into impressive displays. Giant LED screens can display changing images during concerts.

Houses of the future will have LEDs that you can switch on and off, or make dimmer and brighter, by waving your hands or speaking.

▲ LEDs are much more efficient than incandescent lightbulbs.

Key Words

efficient, light-emitting diode (LED), light-dependent resistor (LDR), sensor, sensing circuit

P3 Chapter 1: New technology

How efficient?
LEDs of the future could be 45% efficient. For every 100 J that a future LED uses state:

- how much energy is used to light the room
- how much energy is used to heat it.

Detecting light

Have you ever gone to school and left the light in your bedroom on all day?

A house of the future could have circuits that detect if you are out, if it is daytime, and if the light is on – and then turn it off.

To detect light level you use a component called a **light-dependent resistor** or **LDR**. This is a **sensor**. It detects changes in its surroundings. The resistance of the LDR changes in different light levels.

B Name a component that can detect light level.

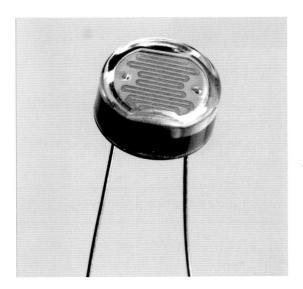
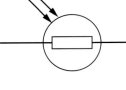

▲ An LDR and its circuit symbol.

If the light level is high the resistance of the LDR is low. To detect if it is daytime you need an LDR outside, and to detect if the light is on you need an LDR in the room. They would be part of a **sensing circuit**.

You can use a sensing circuit to control the light level in a house. It is possible to control circuits in your house by using your phone.

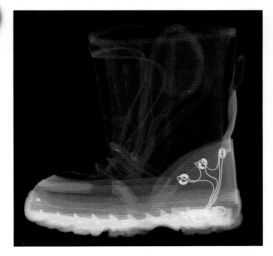

▲ LEDs can be built into clothing or footwear.

Fantastic Fact
If all the energy in your food heated you up you would glow like a 60 W lightbulb.

Summary Questions

1. Copy and complete these sentences.

   A more efficient lightbulb is better at _____ a room than a less efficient lightbulb. You can sense light level with an _____. This could be used in a _____ circuit to control the lights in a house.

   (3 marks)

2. Compare LEDs and incandescent bulbs.

   (2 marks)

3. Describe how an LDR detects light.

   (2 marks)

4. Explain why LEDs could save money compared with incandescent bulbs. Use the efficiency equation to help.

   (6 marks QWC)

1.3 Your hospital – intensive care

Learning objectives
After this topic you will be able to:
- describe how a thermistor detects changes in temperature
- describe how sensors can be used in hospitals.

Many premature babies would not survive without an incubator. Their temperature needs to be kept constant. Equipment in intensive care keeps patients alive.

Monitoring the patient
How hot?
It is very important to keep the temperature inside an incubator high enough so that the baby is kept warm. You can monitor the temperature using a circuit component called a **thermistor**. The resistance of a thermistor changes with temperature. If the temperature increases the resistance decreases. If the temperature decreases the resistance increases.

 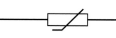

▲ A thermistor and its circuit symbol.

If the temperature inside the incubator falls, the voltage across the resistor increases. This could be used in a sensing circuit to switch on a heater. The thermistor can also be used to make sure that the temperature does not rise too high.

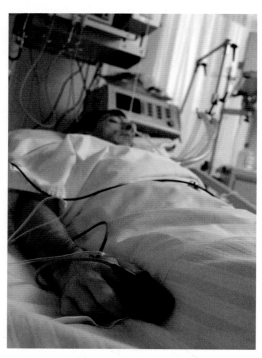

▲ A sensor on the finger monitors pulse rate and oxygen levels.

A State the name of the component that detects temperature.

Pulse rate and oxygen
Your pulse rate shows how fast your heart is beating. Doctors and nurses can take your pulse by feeling with their fingers on your wrist or neck. But in hospital it is often important to monitor the pulse rate continuously, and also the oxygen in the blood.

If you hold a torch under your finger in the dark you will see that your finger is translucent. To monitor the pulse rate and oxygen levels, a sensor on the patient's finger emits visible light and infrared radiation. Blood with lots of oxygen absorbs the infrared more than the visible light. This is how they monitor the oxygen levels.

Key Word
thermistor

Press release
Summarise how physics is used in an intensive care unit in under 100 words.

Heartbeat

For heart patients it is important to monitor electrical changes in the heart, which happen every time the heart beats. Doctors attach electrodes to the patient's skin. These are gel pads that detect small changes in potential difference. The machine amplifies the potential difference and displays it on a screen. The information can be converted to a digital signal that is used by a computer.

B State what electrodes attached to the skin detect.

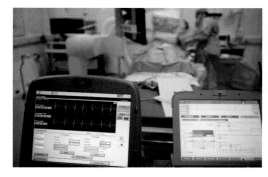

▲ The screen shows the changes in potential difference as the heart beats.

Keeping the patient alive

There are lots of pieces of equipment in an intensive care unit that use physics to help keep people alive.

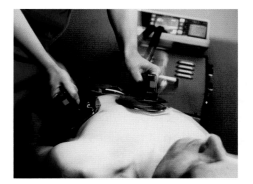

▲ A defibrillator uses stored charge to produce a large potential difference. First aiders use this to start someone's heart when it has stopped beating.

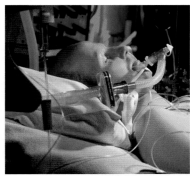

▲ A ventilator uses an electric motor to move a piston up and down. This pumps air into the lungs of a patient who is struggling to breathe.

When your heart beats it produces enough pressure to squirt blood 10 m.

Summary Questions

1. Copy and complete the sentences below.

   You can use a _____ in a circuit to detect temperature. Doctors use visible light and _____ radiation to monitor a patient's pulse. Electrodes connected to a patient's chest can measure _____ _____. A _____ can keep a heart pumping, and a _____ can start a heart that has stopped.

   (5 marks)

2. Describe how you can use a thermistor to monitor temperature in a circuit.

   (3 marks)

3. Compare how pulse rate and heartbeat are monitored.

   (6 marks QWC)

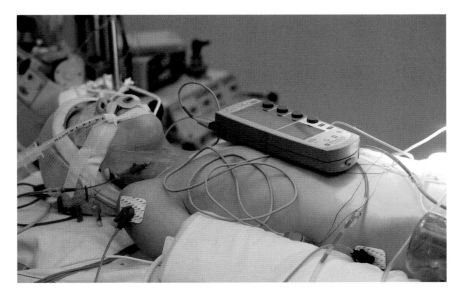

▲ A pacemaker can keep a patient's heart beating. It applies a potential difference at regular intervals to the heart muscles.

1.4 Your hospital – seeing inside

Learning objectives

After this topic you will be able to:
- describe how optical fibres work
- describe some techniques for seeing inside the human body.

You may have had an X-ray, or know someone who has. Before X-rays were used, the only way to find out what was happening inside was to cut a patient open. Today doctors have lots of ways of seeing inside the body.

Using waves of the electromagnetic spectrum
Visible light

One of the simplest ways of seeing inside someone is to use a 'light pipe' called an **endoscope**.

If light hits the boundary between glass and air at a shallow angle it refracts. If it hits at a very large angle it reflects. There is a special angle, called the **critical angle**, where the angle of refraction is 90°.

A State what is meant by critical angle.

This type of reflection, where the light reflects back and stays inside the glass, is called **total internal reflection**. It happens inside thin glass fibres called **optical fibres**.

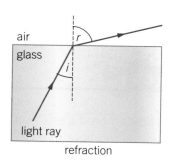
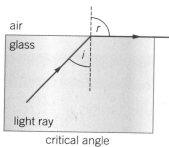
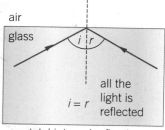

▲ Light can be refracted or reflected inside a piece of glass.

▲ Light is totally internally reflected as it goes down an optical fibre.

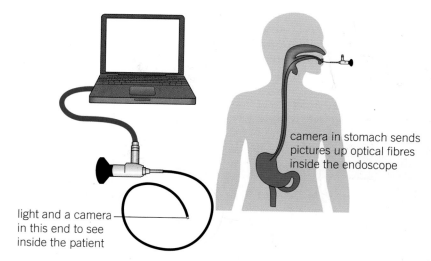

▲ An endoscope contains optical fibres to see inside a patient.

X-rays

X-rays are waves of the electromagnetic spectrum. They have a much higher frequency, and higher energy, than light.

Bones absorb X-rays more than soft tissues such as muscles and internal organs. A machine emits X-rays that travel through your body to a detector. The amount of X-rays that pass straight through or are absorbed builds up a picture of what's inside the body.

● P3 Chapter 1: New technology

Gamma rays

You can see how well someone's kidneys are functioning using **gamma rays**. A patient drinks a medical tracer, a special chemical that is taken in by their kidneys. A detector detects the gamma rays and forms an image. Gamma rays have the highest frequency and energy of all the waves of the electromagnetic spectrum.

B Name three waves of the electromagnetic spectrum that are used to see inside the human body.

Using sound and magnets
Ultrasound

Doctors can use an ultrasound transmitter and receiver on the end of an endoscope. This produces an image of internal organs that you cannot see very easily with X-rays or a normal endoscope.

Magnetic resonance imaging

Magnetic resonance imaging (**MRI**) scanners use very, very powerful electromagnets to make images of the inside of the body. They allow doctors to see how organs, including the brain, work. MRI scanners are very expensive and the patient has to lie very still in a small space while the scan is taken.

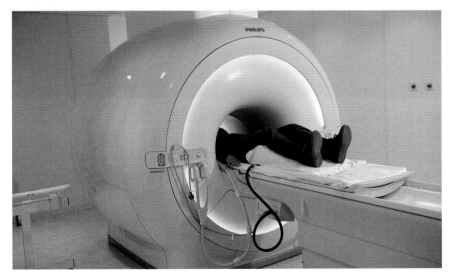

▲ A patient inside an MRI scanner.

Crossword clues

Make a crossword using the waves of the electromagnetic spectrum mentioned on these two pages, as well as ultrasound. Write clues for each type of wave.

Fantastic Fact

The magnetic field inside a hospital scanner is so strong it can levitate a frog. If a magnetic field was strong enough it could levitate a human.

Key Words

endoscope, critical angle, total internal reflection, optical fibre, X-ray, gamma ray, magnetic resonance imaging (MRI)

Summary Questions

1. Copy and complete the sentences below, choosing the correct bold words.

    When light hits the inside of a glass fibre at an angle **bigger/smaller** than the critical angle it is **reflected/refracted**. This is used to see inside patients with an **endoscope/telescope**. Doctors can also use **X-rays/microwaves** and **gamma rays/ultraviolet** to see inside patients.

    (5 marks)

2. Describe the conditions for light to be totally internally reflected.

    (2 marks)

3. Describe how a doctor can investigate the functions of the following organs:
    a the brain   b the kidneys

    (4 marks)

4. Compare the use of light, sound, and magnets to see inside the human body.

    (6 marks QWC)

1.5 Your sports

Learning objectives

After this topic you will be able to:

- describe how technology is used in sport
- describe what is meant by reaction time

▲ Some sports rely on technology.

Key Words

projectile motion, reaction time, uncertainty

▲ It takes time for you to react to what you see or hear.

Sometimes a player challenges a call in a game of tennis. An animation on the big screen shows whether the ball was in or out. How does it work?

Spot the ball

If you hit a ball you can work out where it will go if you know the speed at the start, called the initial speed, and the angle at which you hit it. The way it moves is called **projectile motion**. While the ball is in contact with the racket there is a force on it. This changes the speed and direction of motion of the ball.

When the ball is not in contact with anything there are only two forces acting on it: air resistance and weight. The air resistance reduces the speed and the weight pulls it towards the Earth. The ball slows down and falls down.

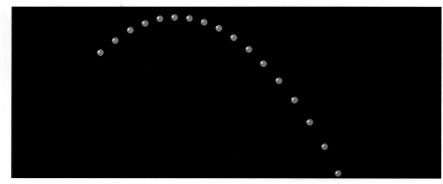

▲ A ball follows a path that you can predict.

A camera can take photographs of the ball many times a second. From this a computer can work out the position of the ball each time and can predict where it will land.

A State the two things that you need to know to work out where a tennis ball will go when you hit it.

How fast can you react?

Think about measuring the time for the 100-m sprint on sports day. You listen for the sound that starts the race and start the stopwatch as soon as you can. It takes time for the signal from your ears to reach your brain, and the signal from your brain to reach your hand. This is called **reaction time**.

B State the name of the time that it takes you to react.

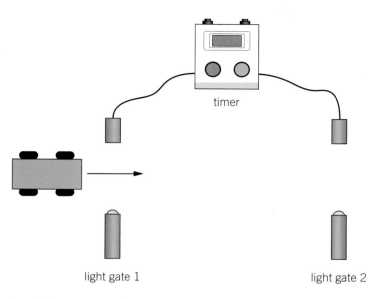

▲ Light gates produce more accurate and precise results than a stopwatch.

▲ Sporting events need to measure the time for the 100-m sprint accurately, without reaction times.

Your reaction time is about 0.2 seconds. You should take this into account when you consider the quality of your data in an investigation. There are more accurate ways of measuring time, such as light gates. In the diagram above, when the car crosses the beam of light at gate 1 the clock starts, and when it crosses the beam of light at gate 2 the clock stops.

Keeping safe

Many sports can be dangerous. Formula 1 drivers wear helmets and their cars are designed to protect them from injury if there is an accident. Formula 1 and other cars have features to increase the time of contact in a collision. If you increase the time that it takes for the car to stop, you reduce the forces on the people inside. They are less likely to be injured.

Uncertainty

If you measure the time for the 100-m sprint your stopwatch might show 00:00:18.3258. This means 0 hours, 0 minutes, and 18.3258 seconds. You should round up to 18.5 seconds because of your reaction time.

There is an **uncertainty** in this result because of your reaction time. Most people's reaction time is about 0.2 seconds.

Summary Questions

1. 🔬 Copy and complete the sentences below.

   A computer can predict where a ball will land because it works out how forces change the _____ and direction of motion of an object. Measuring time with a stopwatch is affected by your _____ time. Racing cars are designed to increase the _____ of a collision to reduce the _____ on the driver.

   (4 marks)

2. 🔬🔬 Describe what is meant by the term reaction time.

   (2 marks)

3. 🔬🔬 Compare the use of a stopwatch and light gates to measure the time it takes a ball to reach the ground.

   (4 marks)

4. 🔬🔬🔬 Discuss the impact that technology has had on sport.

   (6 marks QWC)

1.6 Your planet

Learning objectives

After this topic you will be able to:
- explain why demand for electricity is increasing
- describe how future demand for electricity could be met.

Link

You can learn more about kWh in P2 2.7 Energy and power

Key Words

nuclear fusion

Fantastic Fact

The Sun is getting less massive every day. The energy that it radiates is a result of converting mass to energy.

▲ The UK generates some electricity using wind turbines.

You may have heard people talk about saving energy, or the energy crisis.

What's the problem?

The global demand for electricity is increasing. The chart compares electricity demand between 2009 and 2030, in kWh per person per year. Computer models help scientists to make predictions.

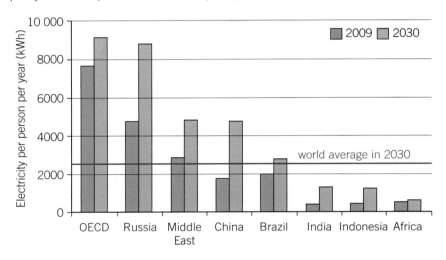

▲ A computer model predicts electricity demand.

Demand could increase for several reasons, such as:
- We use a lot more electrical devices.
- The population of the planet is increasing.
- Countries that did not have a big demand for electricity now have a bigger demand.

There are many consequences of this increase, such as climate change, pollution of the air, water, and soil, and the environmental impact on water and land.

A State one reason why demand for electricity is increasing.

Possible solutions?

We can try to use methods of generating electricity that do not contribute to climate change or produce pollution.

Renewable resources such as wind turbines do not produce carbon dioxide when they are being used. They do produce carbon dioxide when they are being built. Biofuels produce carbon dioxide when they are burned.

Traditional power stations use fossil fuels to boil water to make steam to turn a turbine, which drives a generator. Wind, waves, tides, or falling water can turn a turbine instead. Geothermal energy pumps water underground to produce the steam.

B Name two renewable resources that can turn a turbine.

Solar cells do not need a generator. The cells absorb light from the Sun and produce a potential difference directly. Solar cells are getting more and more efficient. Carbon dioxide is produced when solar cells are made, but not when they are running. In the future solar cells in your clothes could power your phone or other mobile devices.

▲ Solar cells generate electricity in remote places.

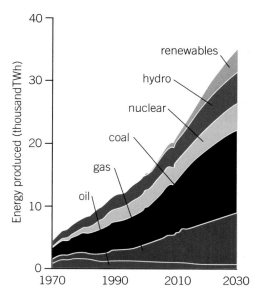

▲ Different resources used worldwide to generate electricity.

Fusion

One of the most exciting projects produces energy from **nuclear fusion**. This is just the same as the process that provides energy in the Sun.

Hydrogen atoms join together to make helium and a lot of energy is produced. The reaction does not produce any greenhouse gases.

▲ A fusion reaction is not like a normal chemical reaction.

C Name the process that makes the Sun shine.

A fusion reactor needs to heat the hydrogen to a very high temperature. If scientists could make a working reactor then our energy problems could be solved.

How many?
Write an electricity journal for a day that lists everything electrical that you use.

Summary Questions

1 Copy the sentences below, choosing the correct bold words.

The demand for electricity is **decreasing/increasing** because the population is **decreasing/increasing**. Another reason is that the number of devices that need electricity is **decreasing/increasing**. We need to find alternatives that **do/don't** contribute to climate change.

(4 marks)

2 Suggest and explain two alternatives to fossil fuels that could generate electricity in the future.

(4 marks)

3 Design a game of snakes and ladders based on the pros and cons of different ways of generating electricity. Write an explanation on each snake or ladder.

(6 marks)

P3 Chapter 1 Summary

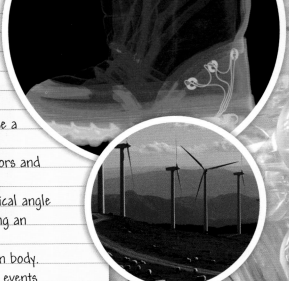

Key Points

- An analogue signal can have any value. A digital signal is a series of 1s and 0s.
- You can use radio waves, microwaves, infrared, and visible light for communication. These waves are part of the electromagnetic spectrum together with ultraviolet, X-rays, and gamma rays.
- A device is efficient if very little energy is wasted. Light-emitting diodes are more efficient than incandescent lightbulbs.
- You can use a light-dependent resistor to monitor light level. You can use a thermistor to monitor temperature.
- You can use sensors to monitor pulse rate and heart rate, and defibrillators and pacemakers to start and maintain a heartbeat.
- Light that hits the inside of a glass fibre at an angle bigger than the critical angle is totally internally reflected. This allows doctors to see inside people using an endoscope.
- Doctors use X-rays, ultrasound, and gamma rays to see inside the human body.
- You can use light gates to reduce the impact of reaction time in sporting events.
- Sports cars are designed to reduce the force on a driver in an accident by increasing the time the collision lasts.
- The demand for electricity is increasing because the population is getting bigger and we are using more electrical devices. In the future the demand will need to be met using a range of renewable and non-renewable resources, or new technology such as nuclear fusion.

Key Words

analogue, digital, binary, bit, electromagnetic spectrum, microwave, radio wave, efficient, light-emitting diode (LED), light-dependent resistor (LDR), sensor, sensing circuit, thermistor, endoscope, critical angle, total internal reflection, optical fibre, X-ray, gamma ray, magnetic resonance imaging (MRI), projectile motion, reaction time, uncertainty, nuclear fusion

Big Write

The big exhibition
Your local council wants to put on an exhibition in a shopping centre that shows the impact of technology on people's lives. They want to make big display boards to highlight how much we all rely on technology without realising it.

Task
Produce designs for at least four boards for the exhibition.

Tips
- Decide which areas of everyday life to include.
- Think about how to introduce the scientific ideas to the general public.
- Make the boards attractive so that people will stop and read them.

End-of-chapter questions

1. State which of the waves in the list below are waves of the electromagnetic spectrum.

   radio waves sound waves visible light
   X-rays infrared water waves
   (4 marks)

2. Match the device to its correct definition.

Device	Definition
LDR	used to see inside people
thermistor	used to detect light
LED	used to detect temperature
endoscope	an efficient lightbulb

   (4 marks)

3. State which of the following times is likely to be your reaction time.

   2 seconds 0.02 seconds
   0.2 seconds 20 seconds
   (1 mark)

4.
   a Name **two** ways that doctors can use technology to see inside the human body without surgery. *(2 marks)*
   b Name **two** ways that engineers can reduce the risk of injury to a car driver. *(2 marks)*
   (4 marks)

5. A student is investigating solar cells. She puts a solar cell on the desk below a lamp and connects it to a voltmeter. She puts thin layers of clear plastic film on top of it to see if the output of the solar cell is affected by them.
   a State the question that she is investigating. *(1 mark)*
   b Draw a table that she could use for her results. *(1 mark)*
   c State and explain the type of graph that she could draw. *(2 marks)*
   (4 marks)

6. The following diagram shows light going into a prism. The critical angle is 42°.

   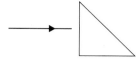

   a Copy and complete the diagram to show what happens to the ray. Use a protractor to measure the angles. *(4 marks)*
   b Explain why you have completed the diagram in the way that you have. *(2 marks)*
   (6 marks)

7. Sketch and explain the shape of a graph of the resistance of a thermistor against temperature. *(2 marks)*

8. Copy and complete the sentences using the words 'digital' or 'analogue'.
   a The signal produced by the microphone in your mobile phone is _____. *(1 mark)*
   b The signal transmitted by your mobile phone is _____. *(1 mark)*
   c The signal received by your friend's mobile phone is _____. *(1 mark)*
   d The signal produced by the loudspeaker in your friend's mobile phone is _____. *(1 mark)*
   (4 marks)

9. Look at the table below. It shows four different devices. Use the law of conservation of energy and the equation for efficiency to complete the table.

Device	Useful energy (J)	Wasted energy (J)	Total energy (J)	Efficiency
lightbulb	5		25	
kettle		500	2000	
television	2500	2500		
car	100		400	

   (8 marks)

10. A gardener wants to use sensors to monitor the temperature in his greenhouse. Describe in detail how he could use thermistors and a heater to keep the temperature in his greenhouse constant.
    (6 marks QWC)

2.1 Discovering the Universe 1

Learning objectives

After this topic you will be able to:
- describe some ideas about the Universe that developed in different cultures
- describe the geocentric model of the Solar System.

How have ideas about the Universe changed? What did people who lived thousands of years ago believe?

Telling stories

People used to tell stories about the Sun, Moon, and other objects in the sky. The stories explained why things happened, like the Sun rising and setting, and why eclipses happened. Some people thought that the Sun was eaten by monsters during a solar eclipse. Many of the ancient ideas about the Earth said that the Earth was flat.

Thousands of years ago there were people who thought that a flat Earth was supported by pillars, or surrounded by oceans. By 300 BC most people in Europe thought the Earth was a sphere, but the flat-Earth idea continued in China until about 1600.

▲ People in Thailand thought that solar eclipses happened when a god called Rahu swallowed the Sun.

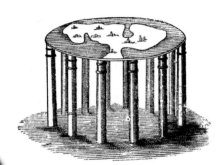

▲ People in India thought that a flat Earth was supported by 12 pillars.

▲ There was another idea, that the Earth was supported on the back of a turtle.

The stories people used to tell were not scientific explanations.

A State two things that scientists do to collect evidence.

What do scientists do?
Scientists use **evidence** from observations and measurements to develop explanations. They use the explanations to make **predictions** about what will happen in the future.

Link
You can learn more about lunar and solar eclipses in P1 4.4 The Moon

Key Words
evidence, prediction, model, geocentric model

Making models

People couldn't look for patterns in their observations of the night sky until writing and numbers had been invented. They needed to make measurements and write them down.

When scientists noticed patterns, they tried to explain them by making **models**. A model can be something physical, or it can use equations. Models help us to explain what we see and to predict what will happen.

P3 Chapter 2: Turning points in physics

Early models of the Universe

Astronomers in India, China, and the Middle East had different ideas about the objects they could see in the sky. Some people believed that the Earth did not move. This was the **geocentric model**. In this model the Sun, Moon, planets, and stars moved around the Earth. They moved on crystal spheres that light could travel through.

Two Greek astronomers, Plato and Aristotle, wrote books explaining this model over 2000 years ago.

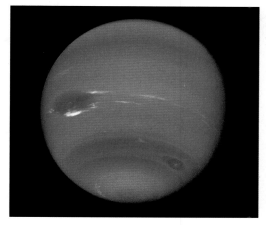

▲ The planet Neptune was where Newton predicted it would be.

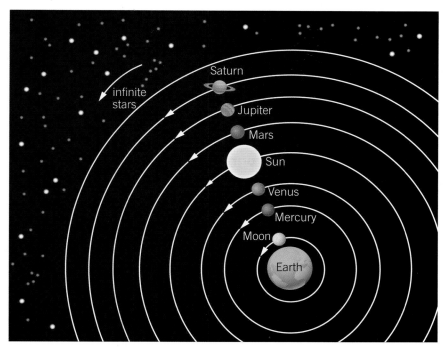

▲ The geocentric model of the Universe with the Earth at the centre.

B Name the model of the Universe with the Earth at the centre.

Same idea, different place

Some ideas are not new.

An Indian mathematician and astronomer called Bhaskaracharya lived around 1000 years ago. He said that there was a force between all objects. This is the force that we call gravity. He worked out that the Earth attracted the Moon 500 years before Isaac Newton published his ideas about gravity.

Newton had the same idea as Bhaskaracharya but he used equations in his explanation. Astronomers used Newton's equations to predict that there was a planet beyond Uranus. Newton's idea was very powerful because you could use it to make predictions.

Strange moon

Imagine you have gone back in time to ancient Thailand. Describe how you could show people how eclipses really happen.

Summary Questions

1 Match the model or story to the country it comes from.

the geocentric model	India
the Earth on the back of a tortoise	Thailand
the Sun is swallowed by a god during an eclipse	Greece

(3 marks)

2 Describe the geocentric model of the Universe.

(2 marks)

3 Think about the stories that people used to tell each other about stars and planets. Compare these with the geocentric model of Newton and Bhaskaracharya.

(6 marks QWC)

2.2 Discovering the Universe 2

Learning objectives

After this topic you will be able to:
- describe how observations led to a different model of the Solar System
- describe the heliocentric model of the Solar System.

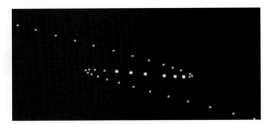

▲ Each dot shows the position of Mars every week for several months.

Key Words

retrograde motion, heliocentric model

You know that the Sun, not the Earth, is at the centre of the Solar System. How did astronomers work that out?

Our model doesn't work

The geocentric model was very popular. Aristotle was a very famous scientist and people trusted what he said. His ideas spread all over the world and people accepted the geocentric model. It explained lots of the observations that people saw every day.

- The ground did not seem to move.
- The Sun and Moon did appear to move.
- The stars also appeared to move.

There was a problem. Sometimes the planets appeared to go backwards. This is called **retrograde motion**. You do not notice it during one night, but if you make observations over several months you can see that the paths of the planets change direction.

A Name the type of motion where the planets appear to move backwards.

It was difficult to explain these observations using the geocentric model. Greek scientists at the time could have rejected the geocentric model and looked for a different model.

They didn't. A scientist called Ptolemy changed the model so that the planets went in complicated orbits. He published his ideas in about 140 AD.

Lots of astronomers all over the world accepted Ptolemy's changes.

A better model

The Greek astronomer Aristarchus had described a different model in around 200 BC. In this model the planets orbited the Sun. This is the **heliocentric model**, the model we use today.

In 1030 Indian astronomers such as Al-Biruni were discussing a heliocentric model.

It was not until 1543 that a book with this model was published. The author was Nicolaus Copernicus.

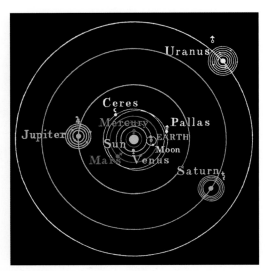

▲ The heliocentric model.

B Name the model of the Solar System with the Sun at the centre.

New technology

Galileo Galilei was an Italian physicist. In 1609 Galileo used a new invention called a telescope. He saw objects in orbit around Jupiter, not in orbit around the Earth. They were moons of Jupiter, and they are called the Galilean satellites after him.

This was evidence that not everything was in orbit around the Earth. This idea went against what the Church believed. Galileo's observations provided new evidence for the heliocentric model, but it took a long time for it to be accepted.

Link

You can learn more about telescopes in P3 3.1 Detecting planets

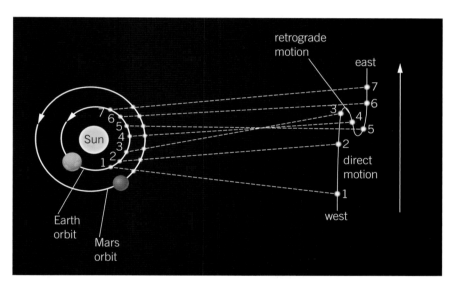

▲ It is easier to explain retrograde motion using the heliocentric model.

Keep it simple

The heliocentric model can explain retrograde motion more simply than the geocentric model. Simple explanations are important in science. If an explanation becomes too complicated scientists will start to look for a simpler explanation.

Date	Where	Idea
850 BC	India	The Earth is in motion around the Sun.
350 BC	Greece	Aristotle describes the geocentric model in a book.
200 BC	Greece	Aristarchus describes the heliocentric model in a book.
151	Greece	Ptolemy publishes his revised geocentric model.
1300	Baghdad	Ibn al-Shatir improves Ptolemy's model.
1543	Poland	Copernicus publishes the heliocentric model.

Foul Fact

You can see one of Galileo's fingers in a museum in Florence, Italy. They were cut off his body 95 years after his death and then lost for nearly 300 years.

Geo-?

'Geo' means 'Earth' in Greek, so 'geocentric' means 'with the Earth at the centre'. Try to work out what the other half of these 'geo' words mean: 'geography', 'geometry'. What do you think 'helio' means?

Summary Questions

1. Copy and complete the sentences below, choosing the correct bold words.

   The geocentric model **did/did not** explain retrograde motion in a simple way. It took observations by **Galileo/Copernicus** to provide evidence for the heliocentric model.
   (2 marks)

2. Explain why Galileo's observations could not be explained using the geocentric model.
   (2 marks)

3. Describe the heliocentric model of the Universe.
   (2 marks)

4. Draw a timeline to show and explain the development of the geocentric and heliocentric models of the Solar System.
   (6 marks)

2.3 The Big Bang

Learning objectives

After this topic you will be able to:
- describe the timescale of the Universe
- describe what is meant by the Big Bang.

How did the Universe begin? When did it start?

Are we special?

When you observe other galaxies you find that they are all moving away from us. You might think that our planet, or our galaxy, is very special, and that we are at the centre of the Universe. This is not the case. If we lived in the Andromeda galaxy we would see exactly the same thing. The Milky Way would appear to be moving away from us.

Imagine that you are on the surface of a balloon that has dots on it. The dots are like galaxies. As you blow up the balloon each dot is moving away from every other dot.

An astronomer called Edwin Hubble made these observations about galaxies moving away from us in 1929. He also found out that galaxies that were further away were moving faster. The Hubble Space Telescope (HST) is named after him.

▲ You can make a model of the expanding Universe with a balloon.

The Big Bang

Most scientists think that the Universe began with the **Big Bang**. All of space and time expanded from something smaller than an atom. The Universe has been expanding ever since. The Big Bang theory explains why galaxies are moving apart and why the galaxies that are further away are moving faster.

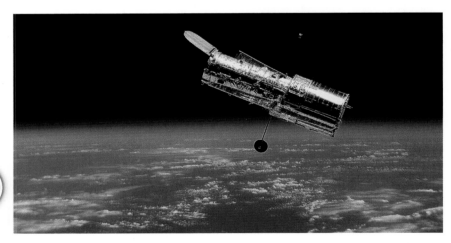

▲ The HST is named after the person who coined the term 'Big Bang'.

A Name the theory that scientists have for how the Universe started.

Just a theory?

Sometimes it is confusing when scientists use the word 'theory'. It doesn't mean that they are not sure. It means that it is the best explanation that scientists have for all of the evidence.

What happened after the Big Bang?

Scientists believe that the Big Bang happened about 14 billion years ago. After about 150 million years the first stars started to appear. Galaxies started to form after billions of years. It wasn't until 9 billion years after the Big Bang that our Solar System formed. That was 5 billion years ago.

◀ Our Solar System formed from a disc of dust and gas.

Life on Earth probably started about 4 billion years ago. The dinosaurs lived between 200 million and 65 million years ago. Humans have existed for less than half a million years.

B State the age of the Solar System.

Modelling timescale

Sometimes it is very difficult to imagine things that are very big, like the timescale of the Universe. It is good to make a model, or an **analogy**, to make it easier to understand.

Imagine that the timescale of the Universe is compressed into a year. If the Big Bang happened on 1 January, the Solar System would form in August, and humans would appear at 11.30 p.m. on 31 December.

The big question

One question that seems very confusing is this one:

"If the Universe is infinite, and it is expanding, what is it expanding into?"

It is hard to think of what an infinite Universe means. It means that there is no edge to it. There is no 'outside' for the Universe to expand into.

Think of the whole of space filled with dough. The dough goes on and on for ever and never stops. Inside the dough are raisins, just like galaxies. The dough expands and the raisins get further apart. This is what is happening in our Universe. The space between galaxies is expanding.

Key Words

Big Bang, analogy

Fantastic Fact

Some scientists believe that microbes on asteroids or comets developed into life on Earth.

▲ The expanding Universe is like expanding dough.

Summary Questions

1. 🧪 Order these events, starting with the earliest.
   first living things
   formation of the Solar System
   Big Bang
   dinosaurs died out
   (4 marks)

2. 🧪🧪 Compare the age of the Earth and the age of the Universe.
   (2 marks)

3. Describe what is meant by the Big Bang.
   (2 marks)

4. 🧪🧪🧪 Draw a scaled timeline of the Universe until today. Include a suitable scale and include key events from these pages.
   (6 marks)

2.4 Spacecraft and satellites

Learning objectives

After this topic you will be able to:
- describe how to get a satellite into orbit
- describe some uses of satellites.

The first living things to go to outer space were fruit flies. How do space scientists launch objects into space?

Into space

You have probably blown up a balloon and let it go. The balloon flies off. The force of the balloon on the air pushes the air backwards. The force of the air on the balloon pushes the balloon forwards. This is a bit like how scientists put a satellite into orbit.

They launch a satellite by attaching it to a rocket. Fuel burns inside the rocket and pushes the waste gases out of the bottom of it. This 'push' is one part of an interaction pair. There is the force of the rocket on the gases, and the force of the gases on the rocket.

There are also forces between the rocket and the Earth. When the force of the air on the rocket is bigger than the force of the Earth on the rocket, it will take off.

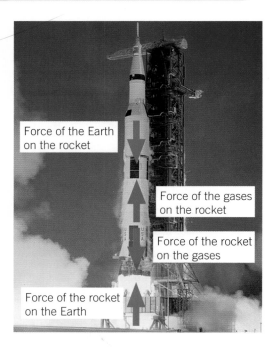

▲ If you ignore drag, there are two forces acting on a rocket when it takes off.

A Name the two forces shown in the image on the left that act on the rocket as it takes off.

There have been more missions to the Moon than to any other object beyond the Earth. That is because the Moon is the closest object, and it only takes days to get there and back.

There have been missions to all the planets in the Solar System and to the Sun. The *New Horizons* spacecraft will take pictures of Pluto and the Kuiper Belt later this decade. Most satellites just take photographs, but missions to Mars have collected rocks and done experiments as well.

▲ The Voyager missions sent spacecraft through the Solar System.

Into orbit

The Moon is the only natural satellite of the Earth. The first artificial satellite was *Sputnik 1*, launched by the Russians in 1957. It was a metal sphere about 50 cm in diameter. To get it into orbit the engineers needed to launch it to the right height and fire it at the right speed. The force of gravity kept it in orbit.

Satellites are used for communication, monitoring the weather, studying the Earth and space, and much more. Scientists need to put the satellite into the right type of orbit for the right job.

Fantastic Fact

Outer space starts 62 miles above the surface of the Earth. Higher than that the air is too thin for planes to fly.

P3 Chapter 2: Turning points in physics

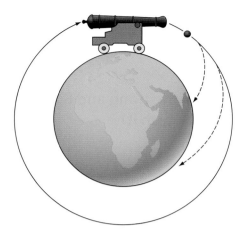

◀ Isaac Newton said that you could launch a cannonball into orbit if you fired it at the right speed.

How many?
The International Space Station takes about 90 minutes to orbit the Earth. Calculate how many times it orbits the Earth each day.

Key Words
geostationary orbit, low Earth orbit (LEO), polar orbit

In the same place
Some satellites stay over the same position on the Earth all the time. They are in a **geostationary orbit**. It takes one day for a geostationary satellite to orbit the Earth. It is always above a spot on the Equator. A geostationary satellite is about 36 000 km from Earth.

A satellite in a geostationary orbit can broadcast television signals. A television company transmits a radio signal to the satellite. The satellite broadcasts it to lots of houses. If you walk down the street you will see that all the satellite dishes point in the same direction.

B Name the type of orbit in which the satellite stays in the same place over the Earth.

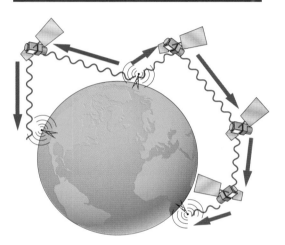

▲ Satellites can relay signals around the world.

Lower orbits
There are lots of satellites, such as the International Space Station, in **low Earth orbit** (**LEO**). This is an orbit below 1000 km from Earth. Some LEO satellites go over the North Pole and the South Pole in a **polar orbit**. Polar orbit satellites are useful for mapping. You need a satellite in LEO if you want to see the whole of the Earth's surface.

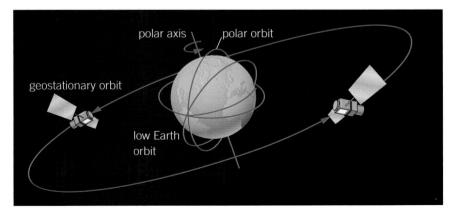

▲ Satellites have different orbits.

Summary Questions

1. Match each satellite to its use.

geostationary orbit	International Space Station
low Earth orbit	satellite television
polar orbit	mapping the Earth's surface

(3 marks)

2. Describe how satellites get into orbit.

(2 marks)

3. Compare geostationary satellites and satellites in low Earth orbit.

(6 marks QWC)

2.5 Mission to the Moon

Learning objectives

After this topic you will be able to:
- describe some of the risks and benefits of the space programme.

▲ Only 12 people have ever seen 'Earthrise' from the Moon.

▲ To land at a safe speed the astronauts needed a parachute to slow them down.

Fantastic Fact

If Buzz had shut the door behind him when he joined Neil on the Moon's surface they would both still be there. There was no handle on the outside of the lander.

It cost a lot of money to send Neil Armstrong and Buzz Aldrin to the Moon in 1969. Was it worth it?

One small step for man ...

The *Apollo* missions from 1961 to 1975 were designed to put people on the Moon. It was the most difficult engineering challenge that had ever been attempted. There are lots of **risks** when you go into space. Equipment can be damaged and people can be injured.

Risk assessment

There is a risk to any activity. It is not possible to make anything 100% safe. When you do a risk assessment you consider the probability that something will happen, and the consequences if it did. Then you do your best to reduce the probability or change the consequences.

In the *Apollo* missions the engineers and scientists tried to make all the equipment 99.9% reliable. There were two or even three back-up systems for each process.

99.9% reliable means that if there were 1000 missions, something would go wrong once. There were only going to be 20 *Apollo* missions to the Moon.

A State the two factors that you need to consider when you do a risk assessment.

A risky business

One of the most obvious risks is at take-off. When the astronauts took off they were lying on top of an enormous rocket that could have exploded. As they travelled back to Earth through the atmosphere the spacecraft got very, very hot due to friction. The outside of the spacecraft was designed to withstand temperatures hotter than the surface of the Sun. To land safely back on Earth they relied on parachutes to slow their descent.

There was danger from the Sun while they were travelling there and back. The Sun produces lots of radiation. Occasionally the Sun emits a **solar flare**. A huge amount of extra radiation is thrown out into space. In 1989 a solar flare wiped out the electricity system in Canada. A solar flare could stop all the on-board computers working.

P3 Chapter 2: Turning points in physics

... one giant leap for mankind

You probably have in your pocket or bag a product of the space programme. In 1963 half the computers in the world were developed for the Moon missions. We would not have smart phones and other computer-controlled gadgets without them. There were lots of other benefits.

The liquid-cooled suits used by racing-car drivers and fire-fighters are based on the *Apollo* astronauts' spacesuits.

The computer programs for swiping credit cards use software designed for the *Apollo* missions.

The shock-absorbing materials used in sports shoes were developed for spacesuits.

Water filters use technology designed to recycle astronauts' urine.

Baby-milk formulas are based on protein-rich drinks developed for astronauts.

One of the most important benefits of the space programme was not a gadget. Neil Armstrong could hold his thumb up and block out the Earth from view. He saw how fragile planet Earth looks from space. This made people think about looking after our planet.

What's the risk?

Suggest one way that a cyclist can reduce the probability of being in an accident.

Suggest one way that a cyclist can reduce the consequences if they are in an accident.

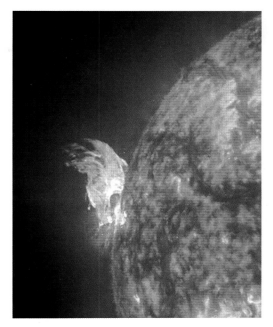
▲ The Sun can give off powerful solar flares.

Key Words

risk, solar flare

Summary Questions

1 Sort each word or words into risks and benefits of the Moon landings.

smart-phone technology
radiation
rocket exploding
baby milk

(4 marks)

2 The *Apollo* space program cost about as much as 5000 hospitals. Describe one benefit of the *Apollo* missions that would be useful in hospitals.

(2 marks)

3 Explain in detail the risks and benefits of the *Apollo* missions to the Moon.

(6 marks QWC)

125

2.6 Radioactivity 1

Learning objectives

After this topic you will be able to:
- describe what is meant by a radioactive material.

What do toys that glow in the dark have to do with radioactivity?

Strange rays

In 1896 a French physicist called Henri Becquerel was interested in rocks that seemed to glow in the dark. You may have seen toys that glow at night after being in sunlight. Becquerel thought that rocks containing the element uranium did the same. He put the rocks in the light, then placed them near photographic plates. These photographic plates were used in cameras before the invention of film and digital cameras. When he developed the plates by processing them with chemicals, he found that the plates had gone dark where the rocks had been near them.

One day it was cloudy so Becquerel decided not to put the rocks in the sunshine. He left them in a drawer instead with some photographic plates. When he processed the plates they had gone dark again. The rocks didn't need to be in sunlight. He thought they must be giving out invisible rays, which he called 'Becquerel rays'.

▲ Some toys glow in the dark after they have been in sunlight.

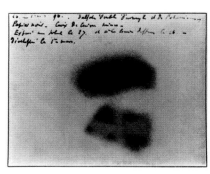

▲ Becquerel's plates had dark patches.

New elements

Marie Curie was a Polish physicist. She was also interested in the rays that rocks seemed to give out. She noticed that some rocks seemed to give out more rays than others. She wondered why.

In 1898 she ground up 1000 kg of a rock called pitchblende. She kept refining it until she had one-tenth of a gram of a new element called **radium**. Radium glows green in the dark. She and her husband Pierre also discovered a new element called polonium.

Marie Curie named the production of invisible rays from some materials '**radioactivity**'. Now we describe these materials as **radioactive**. The mysterious rays are called **radiation**.

▲ Marie Curie.

Fantastic Fact

The cartoon 'The Simpsons' shows radioactive materials glowing green, just like radium.

A Name the word used to describe a material that gives out radiation.

● P3 Chapter 2: Turning points in physics

What are those rays?

You may have met the word 'radiation' in lots of different topics. The Sun gives out radiation. Energy is transferred by radiation. Now you are learning that radioactive materials give out radiation.

Radiation is anything emitted from a source. It can be in the form of a wave or particles. You have already met the waves of the electromagnetic spectrum such as radio waves or infrared radiation. Radiation can also be particles.

Alpha, beta, and gamma

Just a couple of years after Becquerel's discovery another physicist, Ernest Rutherford, worked out that there were three types of radiation emitted by the rocks. Rutherford called them **alpha**, **beta**, and **gamma**. These are the first three letters of the Greek alphabet.

Rutherford and his team of physicists worked out that alpha radiation was made up of the nuclei of helium atoms. Beta radiation was electrons, and gamma radiation was waves of the electromagnetic spectrum.

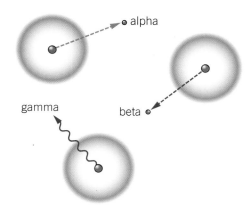

▲ An unstable atom can emit alpha, beta, or gamma radiation.

B Name the three types of radiation given out by radioactive materials.

Half-life

A radioactive material emits 2000 alpha particles per second. If the half-life is 2 days, how much radiation will it emit per second after 4 days?

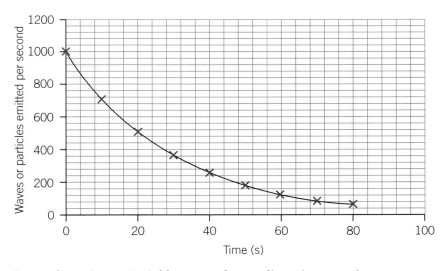

▲ A radioactive material becomes less radioactive over time.

Marie Curie noticed that after a few months her sample of polonium seemed to become less radioactive. All radioactive materials get less radioactive over time. The time it takes for the radiation that they give out per second to go down by half is called the **half-life**. The half-life of polonium is 4.5 months. The half-life of some radioactive materials is millions of years.

Key Words

radium, radioactivity, radioactive, radiation, alpha, beta, gamma, half-life

Summary Questions

1 Copy and complete the sentences below.

Alpha, beta, and gamma radiation is given out by anything that is _____. After a while these materials emit _____ radiation per second.

(2 marks)

2 Describe the difference between 'radiation' and 'radioactive'.

(2 marks)

3 Explain the shape of the graph on this page.

(6 marks QWC)

2.7 Radioactivity 2

Learning objectives

After this topic you will be able to:
- describe the risks of using radioactive materials
- describe some uses of radioactive materials.

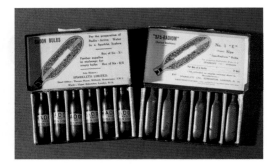

▲ Some people drank a small tube of radioactive water every day.

Key Words

cancer, mutation, sterilise, gamma camera, radiotherapy

How do you know if something is dangerous? If it causes damage that you can see immediately, it will be obvious. If it takes years to see the damage then it is much more difficult.

Miracle cure?

Marie Curie used to keep a small jar of radium by her bed. It glowed in the dark. Her husband carried a small sample in his pocket so that he could show people about the discoveries they had made. People were very excited about this miraculous material.

Very soon people started to sell things that contained radioactive materials, such as radium water and radium toothpaste. They claimed that the radiation could cure disease.

The bad news

Marie Curie and her husband became very ill. At times Pierre's legs were shaking so much he could hardly stand up. In 1932 a famous American golfer died of **cancer** after eating lots of radium for many years. Cancer is a disease caused by cells of the body growing out of control. People had begun to make the connection between radioactive materials and diseases such as cancer.

A Name a disease that radiation can cause.

▲ The hands of this watch glow in the dark because they are painted with radium.

This did not stop people using radioactive materials in a wide range of places. During the Second World War a company that made watch dials employed women to paint radium onto the dials. Years later most of the women developed cancer in their mouths from licking the paintbrushes.

Now we understand the effects of radiation on cells. Radiation can cause changes called **mutations** in cells. These mutations can lead to cancer.

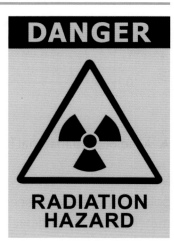

▲ All radioactive materials must be labelled and controlled.

The good news
It is tempting to think that everything about radioactive materials is bad. That is not the case. Radiation can also kill cells such as harmful bacteria. We can use radiation to **sterilise** medical equipment and food.

Doctors also use radioactive materials to diagnose and treat disease.

B State two useful ways that radiation can be used.

Finding out what is wrong
Doctors can use a **gamma camera** to find out whether a patient's kidneys are working properly. The patient drinks a material that is radioactive. The radioactive material emits gamma rays. The material must have a short half-life. Doctors use different types of radioactive material to look at different organs of the body.

▲ This symbol on food packaging shows that radiation has been used to kill bacteria on it.

Fantastic Fact
Some countries such as Japan still sell radium water.

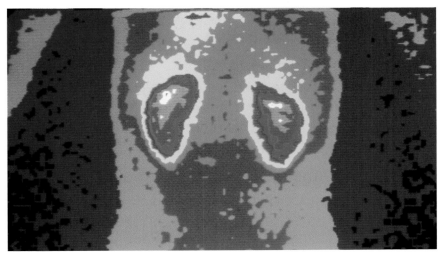

▲ A gamma camera detects radiation.

Treating cancer
You can also use radioactive materials to treat cancers. This might not seem logical because radiation can cause mutations that cause cancer. High doses of radiation can damage cells so badly that they die, and this is how doctors treat cancer cells. This is called **radiotherapy**.

Confusing?
Write a paragraph with as few words as possible that shows the correct use of these words: radioactive, radiation, radiotherapy.

Summary Questions

1. Copy and complete the sentences below.

   People used to think that drinking radioactive water could _____ their diseases. Now we know that radiation can _____ diseases such as cancer. We can also use radioactive materials to cure cancer using _____.

   (3 marks)

2. Describe one risk of using radioactive materials with a long half-life to diagnose disease.

   (2 marks)

3. Compare the risks and benefits of radioactive materials.

   (6 marks, QWC)

2.8 Electromagnetism 1

Learning objectives

After this topic you will be able to:
- describe how to generate electricity using electromagnetic induction.

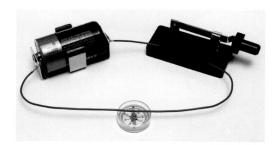

▲ Oersted's experiment showed that electricity and magnetism were linked.

▲ Michael Faraday.

If you rub some materials they can pick up light objects. Some rocks can attract pieces of iron and certain other metals. What is the connection between electricity and magnetism?

A lucky discovery

In 1820 a Danish scientist called Hans Christian Oersted was giving a lecture to other scientists. He wanted to show them some experiments using an electric current to heat a wire. He also wanted to do some experiments with magnetism. He had a compass nearby ready to show the field around a magnet.

He set up the wire and switched the current on. The compass needle moved. He didn't tell the audience, but when they had gone he tried it again and the needle moved again. Oersted had produced a magnetic field around a current flowing in a wire. He had discovered **electromagnetism**.

An important discovery

Michael Faraday was born in 1790 and grew up in London. He left school at the age of 13 and went to work for a person who made books. He started to read the books and learnt a lot of science. Eventually he became a scientist himself.

How do generators work?

Faraday heard about Oersted's discovery and did some experiments himself. He found out that if you move a magnet inside a coil of wire then a potential difference is produced. This is called **electromagnetic induction**. If you connect a meter then a current flows because the circuit is complete.

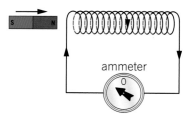

If you move the magnet into the coil the needle moves to the left.

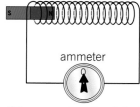

If the magnet and coil don't move the meter reads zero.

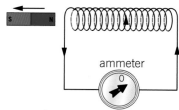

As you move the magnet out of the coil the needle moves to the right.

This is one of the most important discoveries in science. Electricity is generated in a power station by electromagnetic induction. The key thing is that the magnet needs to move relative to the coil of wire.

A Name the process used to generate electricity.

Another scientist, Joseph Henry, discovered electromagnetic induction at exactly the same time. Michael Faraday published his results first so he was given the credit. Scientists publish their results in journals so that other scientists can read them. Before they can be published other scientists check their results. This is called **peer review**.

Backwards and forwards

There is another way to generate an electric current. Instead of using a moving magnet you can use a coil with a current that is changing. This is called **alternating current** and it produces a changing magnetic field. If there is another coil or a piece of metal close by then a current will flow in it.

◀ You can check whether a pan is magnetic with a magnet.

You might have seen a special cooker called an induction hob. There is a coil of copper below the glass hob. It produces a changing magnetic field using alternating current. This generates a current in the bottom of the metal pan, which gets hot and cooks the food. The pan needs to be made of a magnetic material such as steel to work.

Peer review
Describe why it is important for scientists to publish their results in journals, and why peer review is important.

Fantastic Fact
You have used one of Michael Faraday's inventions. He invented an early form of the Bunsen burner.

Key Words
electromagnetism, electromagnetic induction, peer review, alternating current

▲ A wind-up torch contains a magnet and a coil.

Summary Questions

1. 🔬 Copy and complete the sentences below, choosing the correct bold words.

   You can generate a potential difference in a coil of wire by **holding/moving** a magnet in it. If you keep the magnet **moving/stationary** then no potential difference is generated.

   (2 marks)

2. 🔬🔬 Describe how a wind-up torch works in terms of electromagnet induction.

   (3 marks)

3. 🔬🔬🔬 Explain how electricity can be generated in a coil of wire.

   (6 marks QWC)

2.9 Electromagnetism 2

Learning objectives

After this topic you will be able to:
- describe how electromagnetic waves are used for communication.

How do people communicate with submarines, or with satellites? They cannot use wires. The communication must be **wireless**.

From wires to waves

About 150 years ago it took 10 days to send a message from Europe to the United States by ship. Then in 1866 engineers connected the two continents with a transatlantic cable. This meant people could send messages using an electrical signal. The messages were sent as Morse code. This is a series of dots and dashes that represented letters. It was much quicker than a ship, but they could only send 8 words per minute.

There should be waves ...

In 1873 a Scottish scientist called James Clerk Maxwell used ideas from electromagnetism to make a prediction. He said that when electric charges move backwards and forwards such as in an alternating current, it should generate waves. This is a good example of a scientific explanation producing a prediction that you can test. If Maxwell's theory was correct scientists should be able to generate and detect electromagnetic waves.

... and there are

Eight years after Maxwell's prediction, Heinrich Hertz designed some apparatus to test it. His **transmitter** was a circuit with an alternating current that produced a spark. It made radio waves. His detector was a loop of wire. He showed that the waves could travel several metres through the air.

▲ You cannot use wires to communicate with a submarine.

▲ This machine was used to tap out a message in Morse code.

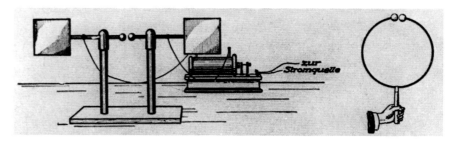

▲ Hertz was the first person to send and receive radio waves.

Fantastic Fact

When someone asked Hertz what his discovery meant for society he said, 'Nothing, I guess'. Little did he know it would lead to WiFi and mobile phones!

You may have seen a loop **aerial** coming out of the back of an old radio or television set. That is how radio waves used to be detected.

● P3 Chapter 2: Turning points in physics

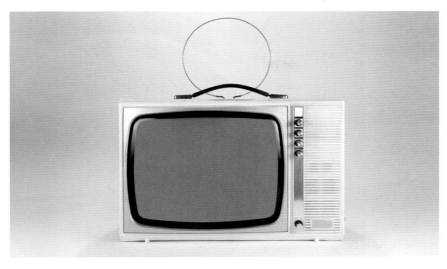

▲ Television signals use electromagnetic waves too.

A Name the piece of equipment that detects radio waves for a radio or television.

Different waves for different uses

How many different wireless devices can you think of? They all work using the electromagnetic waves that Hertz discovered. Waves of different frequencies have different uses. The frequency is the frequency of a radio wave called the carrier wave that the device sends out or receives. Devices add signals that represent sound or images to the carrier waves to send them. An aerial detects the wave and we can see the images and hear the sound.

Frequency range	Example uses
30–300 Hz	communication with submarines
300–3000 Hz	communication within mines
3–30 kHz	navigation, time signals, wireless heart rate monitors
30–300 kHz	AM long-wave broadcasting, amateur radio
300–3000 kHz	AM (medium-wave) radio broadcasts, avalanche beacons
3–30 MHz	citizens' band radio
30–300 MHz	FM radio, television broadcasts
300–3000 MHz	microwave ovens, microwave devices/communications, radio astronomy, mobile phones, WiFi, Bluetooth, GPS
3–30 GHz	radar, communications satellites, satellite television broadcasting

Just a minute …
Calculate how long it would take to send one verse of 'Happy Birthday' along a wire by Morse code.

Link
You can learn more about AM and FM in P3 3.4 Detecting messages

Key Words
wireless, transmitter, aerial

Summary Questions

1. 🧪 Copy and complete the sentences below.

   A _____ wave is a radio wave that carries information such as sound or images. You need _____ communication to talk to people on submarines and aircraft. A device called a _____ produces the waves and an _____ receives them.

   (4 marks)

2. 🧪🧪 Describe how electromagnetic waves are used in communication. Give examples in your answer.

   (4 marks)

3. 🧪🧪🧪 Design a poster to illustrate the information in the table on this page.

   (6 marks)

133

P3 Chapter 2 Summary

Key Points

- Different cultures have developed different ideas about observations of stars and planets.
- The geocentric model was replaced with the heliocentric model as new evidence became available.
- The accepted model for the way the Universe started is the Big Bang.
- Rockets move satellites into orbits that are geostationary orbits, polar orbits, or low Earth orbits. Satellites have a wide range of uses such as for communication and monitoring the weather.
- It was very risky to send people to the Moon but there were many benefits such as advances in computing.
- Some materials give out alpha, beta, and gamma radiation. They are radioactive.
- Radiation can damage cells and cause cancer. It can also be used to sterilise objects and treat cancer in radiotherapy.
- You can generate electricity using electromagnetic induction. A potential difference will be produced if you have a wire in a changing magnetic field.
- Changing electric currents can produce electromagnetic waves. Wireless communication uses different frequencies of electromagnetic radiation.

Key Words

evidence, prediction, model, geocentric model, retrograde motion, heliocentric model, Big Bang, analogy, geostationary orbit, low Earth orbit (LEO), polar orbit, risk, solar flare, radium, radioactivity, radioactive, radiation, alpha, beta, gamma, half-life, cancer, mutation, sterilise, gamma camera, radiotherapy, electromagnetism, electromagnetic induction, peer review, alternating current, wireless, transmitter, aerial

Big Write

The 20th century
There were lots of new discoveries between 1900 and 2000.

Task
Construct a timeline that shows the different discoveries. Use this to write an illustrated story for children.

Tips
- Include what the discoveries were and when they happened.
- Describe how they changed how people lived, or how they thought about the Universe and our place in it.

End-of-chapter questions

1 Match each model to its definition.

geocentric model	The Universe began when all of space and time expanded.
heliocentric model	The Earth is at the centre of the Universe.
Big Bang theory	The Sun is at the centre of the Universe.

(3 marks)

2 State which of these is the approximate age of the Universe in billions of years:

5   14   8   65

(1 mark)

3 From the list below, name the different types of radiation given out by a radioactive material.

alpha   microwaves   sound
gamma   radio waves   beta

(1 mark)

4 Copy and complete these sentences.
 a If a satellite is always over the same place it is in _____ orbit. *(1 mark)*
 b A satellite that goes over the North Pole is in a _____ orbit. *(1 mark)*
 c The International Space Station is in _____ orbit. *(1 mark)*

(3 marks)

5 A student moves a magnet into a coil and the needle on the ammeter in the circuit moves to the left.

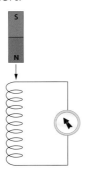

 a Describe what she would see if she moved the magnet out of the coil. *(1 mark)*
 b Describe what she would see if she held the magnet stationary inside the coil. *(1 mark)*

She wants to investigate how the number of turns on the coil affects the size of the current.

 c Describe **two** control variables in this experiment. *(2 marks)*
 d Describe and explain the graph that she could draw using her results. *(2 marks)*

(6 marks)

6
 a Explain how you can use radiation to sterilise needles in hospitals. *(2 marks)*
 b Explain why doctors use materials with a short half-life to diagnose disease. *(2 marks)*

(4 marks)

7 Describe how you can communicate with a submarine when it is underwater.

(2 marks)

8 Describe one piece of evidence that the Universe started with the Big Bang.

(1 mark)

9 You have two pieces of radioactive material. They both give out 400 waves or particles per second.

Sample A has a half-life of 2 days, and Sample B has a half-life of 5 days. Compare how the radiation emitted per second by each sample changes over 10 days.

(4 marks)

10 Explain why doctors need to balance risk and benefit when they use a gamma camera.

(6 marks QWC)

3.1 Detecting planets

Learning objectives

After this topic you will be able to:
- describe how astronomers use telescopes
- describe two types of telescope.

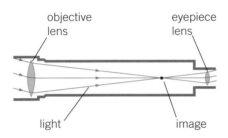

▲ A refracting telescope uses lenses to refract light.

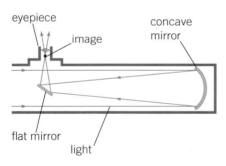

▲ A reflecting telescope uses a concave mirror to reflect light.

Link

You can learn more about lenses in P1 3.3 Refraction

Key Words

refracting telescope, objective, eyepiece, magnify, reflecting telescope, Hubble Space Telescope (HST), exoplanet, intensity

You can see the planets Mercury, Venus, Mars, Jupiter, and Saturn with the naked eye. How do astronomers find planets around stars other than the Sun?

Telescopes

Some of the most important observations in the history of science have been made with the telescope.

Galileo's telescope

Galileo used a telescope with two lenses. This is called a **refracting telescope**. The **objective** lens produces a real image of a distant object. The **eyepiece** lens **magnifies** the image of the object. This is how Galileo could observe the moons in orbit around Jupiter. Before that most people thought that the Earth was at the centre of the Solar System.

A Name the two types of lens that Galileo used in his telescope.

Astronomers observed Neptune for the first time in 1846. They were looking for it because of a prediction from Newton's law of gravitation. This discovery made scientists more confident that Newton was correct.

▲ You can use two lenses to make a telescope.

▲ A satellite dish focuses radiation.

Using curved mirrors

There are problems with Galileo's telescope. For example, different colours of light are refracted by different amounts. So the image of an object that emits different colours might not be clear. Instead of using lenses, it is better to produce an image with a curved mirror. This is called a **reflecting telescope**.

Curved reflectors can be used to focus light or other radiation. This is how a TV satellite dish detects microwave radiation from a satellite.

● P3 Chapter 3: Detection

The Hubble Space Telescope

Curved mirrors can be big, such as the 2.4 m mirror in the **Hubble Space Telescope** (**HST**). It would be difficult to make a lens that big, and it would be too heavy to send into space.

The HST takes stunning images of objects in our Solar System, showing stars being born and stars dying. It has even produced images of planets around other stars, called **exoplanets**.

The light reflected from an exoplanet is very faint. It is difficult to see against the star's light. Astronomers have found thousands of exoplanets. They saw only a few of them with a telescope that uses visible light. They used other methods to find the rest.

B Name the type of planet around a star that is not our Sun.

Kepler

In 2009 the *Kepler* space observatory was launched. It has a telescope with a 95 megapixel camera, the largest launched into space. Its large mirror has detected thousands of exoplanets.

Kepler was launched to try to find planets in other solar systems that might support life. *Kepler* looks at the same part of the sky all the time. If a planet moves in front of a star it will block out some radiation. *Kepler*'s camera can detect the change in the **intensity**, and work out the size of the planet passing in front of the star. The intensity tells you the energy per second transferred as light radiation.

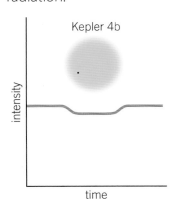
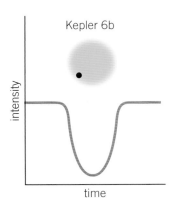

▲ The graphs show the drop in light intensity as a planet moves in front of a star.

Front-page news
Write a headline and paragraph that you might have read in a newspaper when the first exoplanet was discovered in 1992.

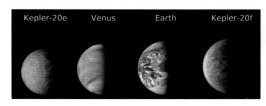

▲ Some of the planets that Kepler has discovered are the same size as Earth.

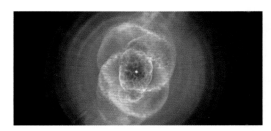

▲ Cat's Eye Nebula image from the HST.

Fantastic Fact

The *Kepler* telescope can detect a change in intensity equivalent to that of a fly in front of a car headlamp on full beam.

Summary Questions

1 🧪 Copy and complete the sentences below.

You can make a telescope with two lenses. The lens near the object is the _____ lens and the lens that you look into is the _____ lens. Astronomers look for planets around other stars, called _____, by detecting changes in the _____ of the light from the star.

(4 marks)

2 🧪🧪 Describe one similarity and one difference between reflecting and refracting telescopes.

(2 marks)

3 🧪🧪🧪 Compare the HST and the *Kepler* space observatory.

(6 marks QWC)

137

3.2 Detecting alien life

Learning objectives

After this topic you will be able to:
- describe how astronomers search for life on other planets.

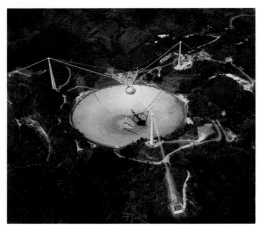

▲ The Aricebo radio telescope looks for radio signals from aliens.

Link

You can learn more about the spectrum of white light in P1 3.5 Colour.

▲ Mars is in the habitable zone. The *Curiosity* rover has found evidence that water flowed on the red planet.

Some people think there is life on other planets. Could there be life, and how would we know?

Is there anyone out there?

There are billions of stars in the billions of galaxies that make up the Universe. Astronomers have detected exoplanets around lots of stars. This makes some astronomers think there may be life on other planets. They estimate that there could be millions or even billions of Earth-like planets in the Milky Way alone.

Scientists use **mathematical models**. These are very useful when you cannot do experiments to collect data. In 1961 the astronomer Frank Drake wanted to show people how scientists work. He wrote an equation for working out how many planets are likely to have life on them. The search for alien life continues.

SETI

SETI is the **Search for Extra-Terrestrial Intelligence**.

Astronomers on the SETI project use radio telescopes, like the one at Aricebo in Puerto Rico, to detect radio signals from space.

Most radio waves, visible light, and some infrared and ultraviolet radiation reach the surface of the Earth. The rest are absorbed by the atmosphere.

Astronomers have put telescopes that detect the other waves of the electromagnetic spectrum in orbit around the Earth.

A State what SETI stands for.

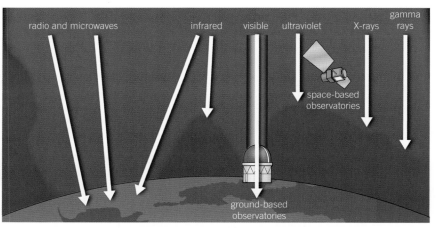

▲ The atmosphere absorbs some waves of the electromagnetic spectrum.

P3 Chapter 3: Detection

Conditions for life

Astronomers use computers to estimate the '**habitable zone**' around other stars. This is the region where an exoplanet would be at a suitable temperature. For there to be life on a planet like Earth there needs to be liquid water, so it cannot be too hot or too cold.

B State what is meant by a habitable zone.

You need an atmosphere with oxygen for life like the life on Earth. Astronomers can work out what stars are made of by looking at the light that they emit. There are black lines, a bit like a barcode, in the spectrum of light. The barcode is called a **line spectrum**. It shows that some frequencies are missing.

▲ The spectrum of the Sun has some frequencies of light missing.

Scientists can do experiments with gases on Earth and work out what the barcodes mean. The lines show which elements are present in the gas that the light has travelled through. That is how we know that the Sun is made mainly of hydrogen and helium.

Astronomers can look at the barcodes in light that has travelled through the atmosphere of an exoplanet as it passes in front of a star.

Can we talk to them?

In 1936 the first radio signal went out into space. If there was a planet with alien life on it around our nearest star, Proxima Centauri, then the aliens would have received the signal in 1940. Proxima Centauri is about 4 light-years away.

Astronomers have sent messages out into space. In 1974 they beamed a message from the Aricebo telescope. It was made of lots of 1s and 0s (bits) that make an image when put together. The image shows numbers, elements, DNA, people, and our place in the Solar System. It was beamed at a star cluster 25 000 light-years away. It will be over 50 000 years before we could get a reply.

▲ The message sent from Aricebo.

Fantastic Fact

Helium was first detected by looking at light from the Sun, not by finding it on Earth.

Key Words

mathematical models, SETI (Search for Extra-Terrestrial Intelligence), habitable zone, line spectrum

How long?

Estimate how long it would take to send a message that is two minutes long to someone on a planet orbiting Proxima Centauri.

Summary Questions

1. Copy and complete the sentences below.

   When they look for alien life astronomers detect _____ waves from space using telescopes on the ground. They look for planets that might have liquid _____ or _____ gas needed for life like the life on Earth.

   (3 marks)

2. Explain why astronomers might not receive signals that aliens send using X-rays.

   (1 mark)

3. Explain three techniques that astronomers use to identify an exoplanet that has the conditions needed for life.

   (6 marks QWC)

3.3 Detecting position

Learning objectives

After this topic you will be able to:
- describe how GPS works
- describe how you can find the distance to planets and stars.

A smart phone can pinpoint its position to within about 10 m. How does it do this?

Using radio waves

You can communicate using radio waves. Your television or radio receives information via radio waves. The **Global Positioning System** (**GPS**) uses 24 satellites in orbit around the Earth. They transmit radio waves.

A State what GPS stands for.

GPS calculates distance from a measurement of time. Radio waves travel at the speed of light, 300 000 000 m/s. If you know the time the waves have taken to travel, and their speed, you can work out the distance they have travelled. This is how GPS works.

▲ Scientists use GPS to track animals such as giant turtles.

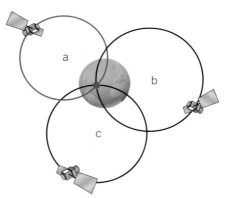

▲ A GPS receiver uses three satellites to work out position.

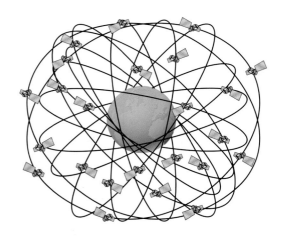

◀ There are GPS satellites that can be used wherever you are on Earth.

A GPS receiver detects signals from three satellites. It works out the distance to each one. From these distances it can work out your position. It may check using a fourth satellite.

GPS signals need **line of sight** to the satellite. There needs to be a direct line from your receiver to the satellite.

Radar

Other systems use radio waves such as **radar** (**RA**dio **D**etection **A**nd **R**anging). Radio waves bounce off metal objects, just as light reflects off a mirror. Radar is similar to sonar. You can find the distance using the time the reflected radio wave takes to travel. For radio waves you need to use the speed of light, not the speed of sound.

▲ It is very important to be able to detect the position of a plane.

Different frequencies have different uses. Air-traffic controllers use radio waves with a frequency of about 3000 MHz to monitor planes. 1 MHz = 1 million Hz. A GPS signal from a satellite has a frequency of about 1575 MHz.

B State what type of waves are used in a radar system.

How do we know?

'How do we know?' is one of the most important questions that you can ask. How do we know the distance to the Moon, or to Mars?

Astronomers find distances to objects in the Solar System using radar. A pulse of radio waves will take about 2.5 seconds to reflect back from the Moon, and over 5.5 hours from Pluto.

Astronomers need a different technique for objects outside the Solar System. This method is called **parallax**. It works because a star such as Proxima Centauri appears to be in a different place in July and in December. Knowing the distance from the Earth to the Sun, astronomers use mathematics to work out the distance to the star.

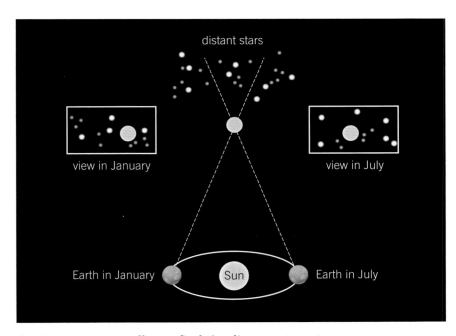

▲ You can use parallax to find the distance to a star.

Fantastic Fact

The clocks on satellites run faster than clocks on the ground. This is an effect predicted by Einstein's theories of relativity. This means that your head is older than your feet!

Link

You can learn more about using different frequencies to broadcast signals in P3 2.9 Electromagnetism 2

How close?

When radio waves travel from a GPS satellite to a receiver the uncertainty in the time is about 0.000 000 03 seconds. Work out the uncertainty in the position using distance = speed x time. (The speed of light is an accurate value.)

Key Words

Global Positioning System (GPS), line of sight, radar, parallax

Summary Questions

1. Copy and complete the sentences below.
   GPS measures the _____ for a signal to reach you from three satellites and works out _____. You can work out the distance to objects in the Solar System using _____. You use a method called _____ to work out the distance to other stars.
   (4 marks)

2. Describe how GPS works.
   (4 marks)

3. Compare the use of radar and parallax for measuring distance.
   (6 marks QWC)

3.4 Detecting messages

Learning objectives
After this topic you will be able to:
- describe how a radio wave carries a signal.

The air is full of waves carrying information. How do devices like radios and mobile phones get the information from the correct wave?

Sending information
Phones, televisions, and radios use waves of the electromagnetic spectrum. Radio waves, microwaves, and infrared are all used as **carrier waves**. The sound that you hear when you listen to a radio station is transmitted using a carrier wave of a particular frequency.

AM and FM radio
When a radio presenter speaks the sound is converted to an electrical signal by the microphone. This electrical signal can then be converted into a digital signal of 1s and 0s.

A radio transmitter adds the digital signal of the sound to the carrier wave. It can change either the amplitude or the frequency of the carrier wave. This is called **modulation**. It produces an AM (amplitude modulated) wave or an FM (frequency modulated) wave.

▲ A microphone converts speech into an electrical signal.

Link
You can learn more about digital and analogue signals in P3 1.1 Your phone

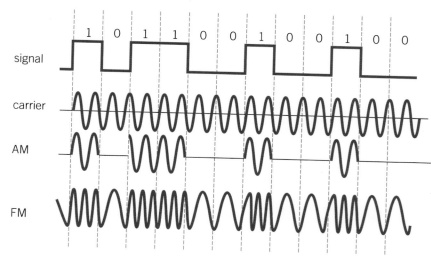

▲ A signal changes either the amplitude or the frequency of a carrier wave.

A Name the two types of modulated wave.

Key Words
carrier wave, modulation, diffraction

The radio transmitter emits the AM or FM carrier wave, and the receiver (your radio) detects the signal. It decodes the signal to get back to sound.

Different radio stations broadcast at different frequencies. The frequency is the frequency of the carrier wave they use. This is the wave that you tune into to listen to your chosen radio station. For Radio 1 it is 99.1 MHz. There are 1000 Hz in 1 kHz, and 1 000 000 Hz in 1 MHz.

Receiving information

Radio waves spread around obstacles like buildings or hills. This is called **diffraction**. All waves can be diffracted.

Diffraction happens when waves go through a gap or around an obstacle. You can see diffracton happening with water waves. They spread out when they go into harbours.

You notice diffraction whenever you can hear someone through a doorway but you cannot see them. The sound waves are diffracted by the doorway but the light waves are not diffracted very much at all.

Link

You can learn more about using different frequencies to broadcast signals in P3 2.9 Electromagnetism 2

How many hertz?
Convert the frequency of Radio 1 to Hz, and to kHz.

Fantastic Fact

Radio waves are the fastest way of sending information because they travel at the speed of light, and nothing travels faster than this.

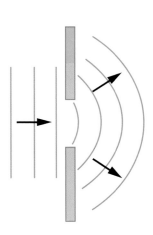

▲ Waves spread out as they go through a gap or round an obstacle.

▲ Water waves diffract as they go into a harbour.

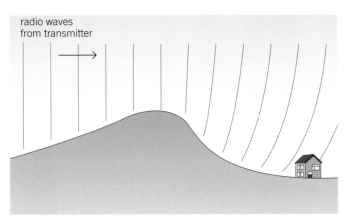

▲ Radio waves diffract around hills.

Summary Questions

1. Match each word to its definition.

diffraction	the radio wave that carries a signal
carrier wave	the spreading out of waves around an obstacle
modulating	changing the amplitude or frequency of a wave

   (3 marks)

2. Describe the difference between AM and FM.

   (2 marks)

3. Describe in detail how the sound of a Radio 1 presenter travels from the microphone to your radio.

   (6 marks QWC)

3.5 Detecting particles

Learning objectives

After this topic you will be able to:
- describe how physicists investigate what the Universe is made of
- describe how particles can be detected.

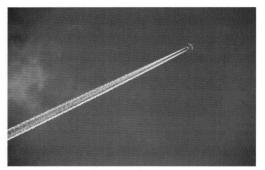

▲ Particles make tracks in cloud chambers just like these tracks.

Link

You can learn more about about the structure of the atom in C3 2.2 Looking into atoms

▲ At the LHC beams are accelerated inside tubes 27 km long in tunnels.

We are made up of tiny particles called atoms. Inside atoms are protons, neutrons, and electrons. What's inside a proton?

Clouds and bubbles

In the past physicists have used cloud chambers or bubble chambers to detect particles that came from space. These detectors produce tracks when particles pass through them, just like the trails made by aircraft in the sky. Physicists could work out the type of particle from the type of track it made. This helps them to work out what the Universe is made of.

In 1936 two American physicists were experimenting with a cloud chamber. They detected a particle that was just like an electron, only more massive. It was a muon. It did not fit with the idea that you needed just protons, neutrons, and electrons to make everything in the Universe. They needed a new model. In the new model there were smaller particles inside protons and neutrons.

A Name one type of detector that produces a particle track.

A smashing time

For over 100 years scientists have smashed things into each other. They have accelerated small particles and fired them into matter. By looking at what came out of the collisions between the particles and the matter, they worked out what is inside matter. This is how Ernest Rutherford worked out that there is a very tiny, massive nucleus at the centre of an atom. This is also how physicists worked out what is inside protons.

New detectors

Physicists accelerate particles such as protons or electrons in a huge machine called an **accelerator**, such as the **Large Hadron Collider (LHC)**. There is a very, very strong electric field inside an accelerator. Protons and electrons move in a field because they are charged particles. This is just like the electrons moving in a wire when you connect up a battery. Connecting the battery produces an electric field. Physicists can accelerate particles in an accelerator to close to the speed of light.

Today physicists don't often use cloud chambers or bubble chambers. The detectors at the LHC are made of **semiconductors** that detect the mass of a particle, and its charge. A semiconductor is half-way between a conductor and an insulator.

B Name the type of material used in particle detectors.

A particle zoo

When physicists smashed beams of protons and electrons together they detected the electrons bouncing off particles much smaller than a proton. This is just like the alpha particles bouncing off the nucleus of an atom. Inside protons (and neutrons) are even tinier particles called **quarks**.

Physicists have detected lots of other types of particle, just as the new model predicted they would. Sometimes the particles that they detect are more massive than the particles that they smash together. This is a bit like smashing an apple and an orange together and getting a watermelon.

The model predicted that there should be a particle called the **Higgs boson**. It was finally discovered at the LHC in 2012, and is one of the last pieces of the jigsaw. The Higgs boson is the particle that gives everything mass.

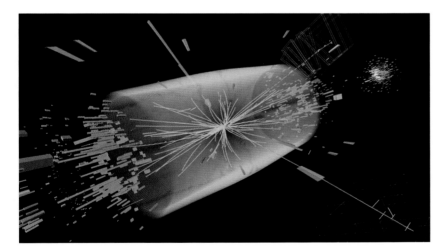

▲ Computers analyse data from the LHC detectors.

Tiny, massive, heavy?
When physicists say that something is massive they do not mean that it is big. They mean that it has mass. Having mass does not mean that something is heavy. Weight is not the same as mass.

▲ A physicist stands in front of one of the huge detectors at the LHC.

Key Words
accelerator, Large Hadron Collider (LHC), semiconductor, quark, Higgs boson

Fantastic Fact
The extra mass made in some collisions comes from the energy of the particles. It is a result of one of the most famous equations in physics: $E = mc^2$. E is energy, m is mass, and c is the speed of light.

Summary Questions

1. Copy and complete the sentences below.

   Physicists used to detect particles from the _____ that they left in cloud chambers. They work out what everything is made of by _____ beams of particles and smashing them together.

   (2 marks)

2. Compare particle detectors used now and in the past.

   (4 marks)

3. Explain in detail the link between the muon, the LHC, and the Higgs boson.

   (6 marks QWC)

P3 Chapter 3 Summary

Key Points

- The first telescope contained an objective lens and an eyepiece lens. Most telescopes are made with curved mirrors.
- Physicists can detect exoplanets by looking at how the intensity of the light radiation from a star changes when the exoplanet passes in front of it.
- We can detect radio signals from space using telescopes on the ground. The Search for Extra-Terrestrial Intelligence (SETI) looks for messages in the signals.
- Physicists can work out which planets are likely to have liquid water. They use the light that has passed through the atmosphere of exoplanets to work out if they contain oxygen.
- The Global Positioning System (GPS) uses the time taken for a radio signal to reach you from different satellites to work out your position.
- Physicists use radar to work out the distance from the time, and this is how they can find the distance to objects in the Solar System.
- Radio waves can carry information such as speech. The sound is converted to a digital signal that is used to modulate a carrier wave (radio wave). Radio waves can be detected behind hills because they diffract around objects and through gaps.
- Physicists accelerate charged particles and smash them together in accelerators like the Large Hadron Collider (LHC). This has helped them to work out that protons and neutrons contain smaller particles called quarks.
- They detect particles from the tracks in cloud chambers or bubble chambers, or by using semiconductors.

Big Write

Cosmic detective
You can see some things that are far away, like planets, with a telescope. You can see things that are really small, like cells, with a microscope, but you can't use a microscope to 'see' tiny particles like protons.

Task
You are a writer for a publishing company. Design a four-page children's leaflet that explains how physicists see things that are far away or very small.

Tips
- Remember to explain clearly each key word you use.
- Include images to make the science interesting for your audience.

Key Words

refracting telescope, objective, eyepiece, magnify, reflecting telescope, Hubble Space Telescope (HST), exoplanet, intensity, mathematical model, SETI (Search for Extra-Terrestrial Intelligence), habitable zone, line spectrum, Global Positioning System (GPS), line of sight, radar, parallax, carrier wave, modulation, diffraction, accelerator, Large Hadron Collider (LHC), semiconductor, quark, Higgs boson

End-of-chapter questions

1. Match the object being detected to the method of detection.

exoplanet	radar
Higgs boson	Kepler Space Telescope
position of a plane	Large Hadron Collider

   (3 marks)

2. Look at the diagrams below.

   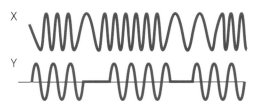

   a State the letter of the diagram that shows an AM wave. *(1 mark)*
   b State what the letter A in AM stands for. *(1 mark)*
   (2 marks)

3. List these objects in order of size, starting with the smallest.

   proton   quark   atom   molecule
   (1 mark)

4. Describe the difference between a planet and an exoplanet. *(2 marks)*

5. Look at the diagram of the telescope.

   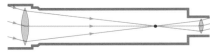

   a Label the eyepiece and the objective lenses. *(1 mark)*
   b State **two** reasons why telescopes in space use mirrors, not lenses. *(2 marks)*
   (3 marks)

6. A radio station broadcasts at a frequency of 93.2 MHz.
   a Convert the frequency to Hz. *(1 mark)*
   b Describe the difference between a sound wave and a carrier wave. *(2 marks)*
   (3 marks)

7. Describe the difference between the detector used to discover the muon and the detector that found the Higgs boson. *(4 marks)*

8. Here is some data about five exoplanets that show the percentage change in the intensity of radiation detected by the *Kepler* observatory as each moves across a star.

   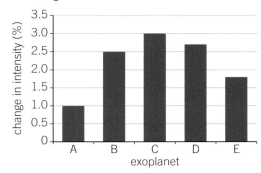

   a List the stars in order starting with the biggest. *(1 mark)*
   b Explain why you have put the planets in the order that you have chosen. *(2 marks)*
   (3 marks)

9. The Chandra telescope detects X-rays, not visible light. Explain why it was sent into space, rather than used on the ground. *(2 marks)*

10.
    a A radar system detects a plane 3 km away. Calculate how long it takes the radio signal to hit the plane and come back. The speed of light is 300 000 000 m/s. *(3 marks)*
    b It takes 2.5 seconds to detect a radio signal reflected from the Moon. Calculate the distance to the Moon. *(3 marks)*
    c State **two** reasons why there is a limit to the distances you can measure using radar. *(2 marks)*
    (7 marks)

11. You are an alien astronomer living on an exoplanet. Explain in detail how you could work out that there was life on Earth. *(6 marks QWC)*

Glossary

accelerator A machine that uses electric fields to make charged particles travel very fast.

aerial A piece of metal used to detect electromagnetic waves.

allele Different forms of a gene.

alpha (radiation) A type of radiation given out by radioactive material; alpha radiation is helium nuclei.

alternating current Current carried by charges moving backwards and forwards.

analogue Can have any value.

analogy A comparison between two things that helps you to explain what one of them is.

antibiotic A medicinal drug that kills bacteria.

antibiotic resistant Bacteria that cannot be killed by an antibiotic.

antibody Chemicals produced by the body that destroy pathogens.

asexual reproduction Reproduction using only one parent.

atom The smallest part of an element that can exist.

beta (radiation) A type of radiation given out by radioactive material; beta radiation is electrons.

Big Bang The expansion of space, which scientists believe is how the Universe started.

binary Can only have two values, 0 or 1.

biodegradable A substance is biodegradable if it can be broken down by natural processes.

biofuel A fuel made from plants or from animal waste.

biotechnology The use of biological processes or organisms to create useful products.

bit A 1 or a 0, from 'binary digit'.

blood group Category of blood, determined by antigens found on the surface of red blood cells.

cancer A disease caused by mutations in cells.

captive breeding Breeding animals in human-controlled environments.

carbon nanotube A cylinder of carbon atoms, often with walls just one atom thick.

carbon neutral A fuel is said to be carbon neutral if the amount of carbon dioxide it takes in when it grows is equal to the amount of carbon dioxide produced when it burns.

carrier A person who has one copy of a dominant allele, and one copy of a recessive allele. A carrier only has the characterisitcs of the dominant allele.

carrier wave A wave, such as a radio wave, that carries information from a transmitter to a receiver.

catalyst Substance that speeds up a reaction without being used up.

catalytic converter A part of a car between the engine and exhaust pipe that converts harmful sustances made in the engine into less harmful ones.

chromatography A technique to separate mixtures of liquids that are soluble in the same solvent.

clone An organism that is genetically identical to its parent.

compound A substance made up of atoms of two or more elements, strongly joined together.

conservation Protecting a natural environment, to ensure that habitats are not lost.

critical angle The angle of incidence for which the angle of refraction is 90°.

denatured The shape of an enzyme is changed if the temperature of a reaction is too high.

dental record A person's dental history used for identification.

diffraction Spreading out of waves around obstacles.

digital Can only have a high or a low value.

DNA Chemical that contains all the information needed to make an organism.

DNA fingerprinting The analysis of DNA from body samples. The DNA fingerprint produced is unique to an individual.

dominant A dominant allele will always be expressed if it is present.

efficient Does not waste much energy.

electromagnetic induction The process of generating electricity using a changing magnetic field and a wire.

electromagnetic spectrum The range of wavelengths of radiation produced by the Sun and other sources.

electromagnetism The interaction of magnetism and electricity.

electron A tiny particle with a negative charge that is part of an atom.

element A substance that cannot be broken down into other substances.

endangered species A species with only small numbers of organisms left in the world.

endoscope A medical instrument for seeing inside the human body using total internal reflection.

evaporate The change of state from liquid to gas that occurs when particles leave the surface of the liquid only. It can happen at any temperature.

evidence Measurements or observations that scientists use to develop or check theories.

exoplanet A planet orbiting a star that is not our Sun.

eyepiece (lens) The lens that magnifies the image in a refracting telescope.

fermentation A type of anaerobic respiration where glucose is converted into ethanol, carbon dioxide, and energy.

filtrate The liquid or solution that collects in the container after the mixture has passed through the filter paper.

filtration A way of separating pieces of solid that are mixed with a liquid or solution by pouring through filter paper.

fingerprint An impression of the ridges present on the surface of your fingers.

forensic science Study of evidence and objects that relate to a crime.

fossil The remains, or traces, of an animal or plant that lived many years ago. Fossils are preserved by natural processes.

gamma camera A camera that detects gamma radiation.

gamma (radiation) A type of radiation given out by radioactive material; gamma radiation is an electromagnetic wave.

gamma ray Wave with the highest frequency in the electromagnetic spectrum.

gas chromatogram An image obtained from gas chromatography.

gas chromatography A technique to separate and identify substances in mixtures, and measure their amounts. In gas chromatography the stationary phase is a polymer, and the mobile phase is helium gas.

genetically inherited disorder A condition passed from parents to their offspring in their genes.

genetic engineering A technique where scientists insert foreign genes into organisms to change their characteristics.

geocentric model A model of the Solar System with the Earth at the centre.

geostationary orbit A satellite in this orbit stays over the same place on the Earth's surface.

Global Positioning System (GPS) The system that uses radio waves from satellites to pinpoint your position.

habitable zone The distances from a star where a planet could have conditions for life.

half-life The time it takes for the amount of radiation per second emitted by a radioactive material to go down by half.

hazard A possible source of danger.

heliocentric model A model of the Solar System with the Sun at the centre.

Higgs boson The particle that gives everything mass.

Hubble Space Telescope (HST) A telescope in orbit around the Earth.

hybrid electric car A hybrid electric car includes an internal combustion engine and a big battery.

hydrocarbon A compound whose molecules are made up of atoms of carbon and hydrogen only.

hydrogen fuel cell An electric cell that generates electricity when hydrogen and oxygen react together.

immune Resistant to a disease.

immune system The body system responsible for fighting disease.

immunisation A method of inserting a vaccine into the body.

index fossil A fossil type that indentifies the geological time period in which a rock was formed.

intensity (of light) Energy transferred per second as light.

Large Hadron Collider (LHC) The particle accelerator that detected the Higgs boson.

light-dependent resistor (LDR) A device that has a resistance that changes with light level.

light-emitting diode (LED) An efficient component that gives out light.

line of sight A direct line between a source and a detector.

line spectrum A spectrum of light that has some frequencies of light missing.

litmus An indicator. Blue litmus paper goes red on adding acid. Red litmus goes blue on adding alkali.

low Earth orbit (LEO) An orbit of about 1000 km above the Earth's surface.

magnetic resonance imaging (MRI) A technique for producing images of the inside of the human body using magnetism.

magnification How many times bigger the image appears compared to the object.

magnify To make something appear bigger.

mathematical model A set of rules using maths, usually using a computer, that helps you predict what could happen.

metal Elements on the left of the stepped line of the Periodic Table. Most elements are metals. They are good conductors of energy and electricity.

microwave Wave of the electromagnetic spectrum used for heating and for communicating.

mobile phase The solvent that carries substances up a piece of chromatography paper, or the gas that carries substances through a gas chromatography column.

model Objects, equations, or images that help to explain what we see.

modulation A process of adding information to a carrier wave.

mutation Change to cells that can cause disease.

nanomedicine The use of nanoparticles to treat disease.

nanometre A unit of length, which is one billionth of a metre (0.000 000 001 m). Its symbol is nm.

nanoparticle Tiny pieces of a substance, with a diamater of between 1 nanometre and 100 nanometres.

nuclear fusion A process that produces energy by combining light nuclei to form heavier nuclei.

objective (lens) The lens in a telescope that focuses light from an object.

optical fibre Very fine tube of plastic or glass that can transmit light using total internal reflection.

parallax A method of working out the distance to stars outside the Solar System.

particulate Tiny pieces of solid, about 100 nm across. Particulates form when diesel burns.

pasteurised Heating a food or drink product to a high temperature to kill any microorganisms present.

pathogen A microorganism that causes a disease.

pathologist A doctor who specialises in understanding the nature and cause of disease.

peer review The evaluation of a scientist's work by another scientist.

Periodic Table A table of all the elements, in which elements with similar properties are grouped together.

plasma Liquid that transports blood cells and other materials around the body.

platelet Fragment of cells that helps the blood to clot.

polar orbit A satellite in this orbit goes over the North and South Poles.

prediction A statement that says what you think will happen.

primate A group of mammals including monkeys, apes, tarsiers, and humans.

projectile motion Motion that can be predicted from the speed and direction of an object thrown into the air.

Punnett square A diagram used to show the possible combinations of alleles inherited from the parents.

quark A tiny particle that protons, neutrons and other particles are made of.

radar RAdio Detection And Ranging, a method of working out distance from the time of reflection.

radiation Waves or particles emitted by a source (which could be radioactive).

radioactive Giving out alpha, beta, or gamma radiation.

radioactivity The name of the process of producing radiation invented by Marie Curie.

radiotherapy A type of therapy that uses radioactive material to kill cancer cells.

radio waves Waves with the lowest frequency in the electromagnetic spectrum, used for communicating.

radium   A radioactive element discovered by Marie Curie.

reaction time   The time the brain takes to process information and act.

reactivity series   A list of metals in order of how vigorously they react.

recessive   A recessive allele will only be expressed if two copies are present.

reflecting telescope   A telescope that uses mirrors to produce an image.

refracting telescope   A telescope that uses lenses to produce an image.

renewable   A fuel that is easily replaced.

resolution   How clearly a microscope can distinguish two separate points.

retrograde motion   The motion of a planet that looks as if it is moving backwards.

rigor mortis   The stiffening of the body a few hours after death.

risk   The probability of something happening that could cause damage or injury.

rust   Hydrated iron oxide, which forms when iron reacts with oxygen and water.

safety   Making sure that something is safe and that hazards and risks are minimal.

scientific journal   A collection of articles written by scientists about their research.

seed bank   A store of genetic material from which new plants can be grown in the future.

selective breeding   Breeding organisms to produce offspring with the desired characteristics.

semiconductor   A material that is halfway between a conductor and an insulator, used to detect particles.

sensing circuit   A circuit that uses components such as LDRs and thermistors to detect changes in the environment, such as light or temperature changes.

sensor   An electrical component that produces a voltage or changes its resistance when it detects something in the environment, such as a change in temperature.

SETI (Search for Extra-Terrestrial Intelligence)   The programme of searching radio signals from space to find evidence of life on other planets.

solar flare   A process that produces a large amount of radiation emitted by the Sun.

stationary phase   The medium that the mobile phase passes through in chromatography. In paper chromatography, the stationary phase is the paper.

sterilise   To kill the bacteria or other microorganisms on an object.

strata   Layers of sedimentary rock.

superbug   Bacteria that are resistant to most types of antibiotic.

thermistor   A device that has a resistance that changes with temperature.

total internal reflection   The complete reflection of light at a boundary between two media.

transmitter   An object that produces (electromagnetic) radiation.

uncertainty The amount by which you cannot be sure of the value of your measurement because of your measuring instruments or methods.

universal indicator An indicator that changes colour to show the pH of a solution. It is a mixture of dyes.

vaccine A substance containing dead or inactive microorganisms used to immunise against disease.

wireless The process of transmitting information using electromagnetic waves rather than wires.

X-ray Wave of the electromagnetic spectrum used for producing images of bones and tissue.

Index

acid rain 63
adenine (A) DNA base 30
aerials 132, 133
Aldrin, Buzz 124
alien life (detection) 138, 139
    conditions for life 139
    DNA 139
    habitable zone 139
    line spectrum 139
    mathematical model 138
    Proxima Centauri 139
    SETI 138
alleles 4
alpha rays 127
alternating current 131
AM and FM radio 142, 143
amplitude modulation (AM) 142
analogy (Big Bang) 121
Andromeda galaxy 120
animal cloning 13
    cell cloning 13
    tissue culture 13
Anning, Mary 79
antibiotics 26–29
    bandages 58
    Fleming, Alexander 26, 27
    MRSA 28
    mutation 28
    penicillin 26–8
    resistant 28
    sterilisation 29
    superbugs 28, 29
    treatment 2
    yoghurt 17
antibodies 24
arch (fingerprinting) 41
Archicebus 80, 81
Aricebo Telescope 139
Aristotle 117, 118
Armstrong, Neil 124, 125
asexual reproduction 12
asthma 63
atoms 72–75, 144
    compounds 73
    Dalton, John 72–74
    electrons 72
    elements 73
    Lavoisier, Antoine 73
    plum–pudding model 72, 73
    Rutherford, Ernest 75

baby food 18
baby-milk 125
bacteria 33
    antibiotic resistant 33
    antibiotics 26–29, 33
    cheese 16
    cloning 12
    fossils 78
    genetic engineering 11
    insulin 12
    MRSA 28
    nanoparticles 54, 58
    penicillin 26, 27
    yoghurt 17
bandages 58
batteries 69
Becquerel, Henri 126
beer 15
Berg, Otto 77
beta rays 127
Bhaskaracharya 117
Big Bang 120, 121
    analogy 121
    Andromeda galaxy 120
    galaxies 121
    Hubble, Edward 121
    Hubble Space Telescope 120
    Milky Way 120
    Solar System 121
    theory 120
    timescale 121
binary numbers 102
biodegradable materials 95
biofuels 65, 113
biotechnology 14–17
    beer 15
    bread 14
    cheese 16
    fermentation 14, 15
    lactose 16
    microorganisms 14
    pasteurisation 17
    sugar 15
    wine 15
    yeast 14, 15
    yoghurt 17
bit (binary digit) 102
bite marks 49
blood
    alcohol 90, 91
    gas chromatography 90, 91
    pathology 48

blood typing 44, 45
    analysis 44
    blood donation 44
    cells 44, 45
    groups 44, 45
    plasma 44
    platelets 44
    transfusions 45
blood/sand separation 85
blowflies 47
body
    appearance 47
    temperature 47
bread 14
butanoic acid 86

cancer 58, 59
    nanomedicine 58, 59
    radioactivity 128
    treatment 129
Cannizaro, Stanislao 76
captive breeding 34, 35
carbon dioxide 14
    biofuels 113
    catalytic converters 66
    fermentation 14
    nanoparticles 54
    new fuels 64, 65
    solar cells 113
carbon nanotubes 56, 57
carbon neutral (biofuels) 65
carpet 94
carrier waves 142
cars 62, 63
    acid rain 63
    hydrocarbons 62
    oxides of nitrogen 62
    particulates 63
catalysts 66, 67
    palladium 67
    platinum 67
    rhodium 67
    rhenium 77
catalytic converters 66, 67
CDs 103
cheese 16
chocolate 72
chromatography 88, 89
    mobile phase 89
    stationary phase 89
cloning 12, 13
    animals 13

asexual reproduction 12
bacteria 12
Dolly the sheep 13
plants 13
rooting powder 13
spider plants 12
clothing fibres 39
cockroaches 47
compounds 73
computer programs 125
conservation 34
corn 10
cotton 10, 39
cowpox 23
Crick, Francis 31
critical angle 108
crude oil 64
Curie, Marie 126, 128
Curie, Pierre 126, 128
cystic fibrosis 6
cytosine (C) 30

dairy cows 8
Dalton, John 72–74
Darwin, Charles 32, 33
    finches 32, 33
    Galapagos Islands 32, 33
    natural selection 32, 33
    peer review 33
    theory of evolution 32, 33
defibrillators 107
Democritus 72
denaturation 19
dental records 96
deoxyribonucleic acid (DNA) 30, 31, 42, 43, 84, 85, 88
diesel 61, 63, 64
    exhaust fumes 61
    fossil fuel 64
    particulates 63
dinosaurs 34
DNA (deoxyribonucleic acid) 30, 31
    alien life 139
    bases 30
    fingerprinting 42, 43
    Franklin, Rosalind 31
    genes 30
    helical structure 30, 31
DNA fingerprinting 42, 43
dogs 9

Drake, Frank 138
DVDs 103

efficiency 104
electromagnetic induction 130
electromagnetic spectrum 103, 108
electromagnetism 130–133
    aerials 132, 133
    alternating current 131
    electromagnetic induction 130
    induction hobs 131
    Morse code 132
    waves 133
electron microscopes 38, 39
electrons 72, 73, 144
elements 73
endangered species 34
endoscopes 108
enzymes 18, 19
    baby food 18
    denaturation 19
    fruit juice 18
    stain removal 18
    temperature 19
    washing powder 18
evaporation 85
evolution theory 32, 33
extinction 34, 35
    captive breeding 34, 35
    conservation 34
    dinosaurs 34
    endangered species 34
    giant pandas 34
    seed banks 34
    woolly mammoths 34

fabrics 57
Faraday, Michael 130
feces 48, 78
fermentation
    equation 14
    sugar 15
    yeast 14, 15
filtration 85
fingerprinting 40, 41
fish that glow in the dark 10
Fitzroy, Robert 32
Fleming, Alexander 26, 27
fly larvae (maggots) 47
forensic science 38, 95
Formula 1 111
fossils 78, 79
    index fossils 79
    rocks 79
    strata 79
Franklin, Rosalind 31
frequency modulation (FM) 142
frost-resistant tomatoes 11
fruit juice 18

Galapagos Islands 32, 33
galaxies 120, 121
Galileo, Galillei 119, 136
gamma camera 129
gamma rays 127
gas chromatography 90, 91
generators 130
genetic counselling 43
genetic disorders 43
genetic engineering 10, 11
genetics 4, 5
    alleles 4
    cloning 12, 13
    engineering 10, 11
    genes 2
    inherited disorders 2, 4, 5
    plant genes 2
    Punnett square 5
    selective breeding 8, 9
geostationary orbit 122
geothermal energy 113
giant pandas 44
global positioning system (GPS) 140, 141
gold nanoparticles 54, 55
GPS (global positioning system) 140, 141
gravity 117
greenhouse gases 65
guanine (G) 30

habitable zone 139
half-life (radioactivity) 127
hazard symbol 87
heartbeat 107
heliocentric model 119
Hertz, Heinrich 132, 133
Higgs boson 145
hospitals 106–109
    defibrillators 107
    premature babies 106
    thermistors 106
    ventilators 107
    endoscopes 108
    magnetic resonance imaging 109
    total internal reflection 108
    ultrasound 109
    X-rays 108
houses 104, 105
    light-dependent resistors 105
    sensing circuit 105
HST (Hubble Space Telescope) 120, 137
Hubble, Edward 121
Hubble Space Telescope (HST) 120, 137
human evolution 81
human hair 55
hybrid electric cars 68, 69
    batteries 69
    fuel 69
hydrocarbons 62
hydrogen fuel cells 64

immune system 22
immunisation 2, 22, 24
immunity 24
index fossils 79
induction hobs 131
inherited disorders 6, 7
    carriers 6
    cystic fibrosis 6
    genetics 2, 6
    polydactyly 7
insects 47
insulin
    cloning 12
    genetic engineering 11

Jenner, Edward 23

Kepler 137

Lactobacillus bulgaricus 17
lactose 16
Large Hadron Collider (LHC) 144, 145
Lavoisier, Antoine 73
LED (light-emitting diode) 104
LEO (low Earth orbit) 123
Leucipus 72
LHC (Large Hadron Collider) 144, 145
light gates 111
light-dependent resistor (LDR) 104, 105
light-emitting diode (LED) 104
line spectrum 139
liquid-cooled suits 125
live yoghurt 17
loop (fingerprinting) 41
low Earth orbit (LEO) 123
lower orbits 123
LDR (light-dependent resistor) 104, 105

magnetic resonance imaging (MRI) 109
magnets 109
magnification 38
maternity/paternity testing 43
Maxwell, James Clerk 132
measles, mumps, and rubella (MMR) 25
medicine
    nanoparticles 58, 59
    nanotechnology 3
Mendeelev, Dmitri 76
meningitis 27
microorganisms 14, 22, 29
microscopy 38, 39
    clothing fibres 39
    electron microscopes 38, 39
    forensic science 38
    light microscops 39
    magnification 38
    resolution 38
microwaves 103
Milky Way 120, 138
Millennium Seed Bank 35
mites 47
MMR 25
mobile phase (chromatography) 89
mobile phones 100
modulation 142
Moon 122, 124, 125
Morse code 132
MRI (magnetic resonance imaging) 109
MRSA 28
mutation 28

nanomedicine 58, 59
    bandages 58
nanoparticles 52, 54–61
    bullet-proof vests 54
    carbon nanotubes 56, 57
    diesel exhaust fumes 61
    fabric protection 57
    gold 54, 55
    human hair 55
    medicine 58, 59
    properties 54
    safety 60, 61
    sunscreen 60, 61

natural selection 32, 33
neutrons 144
new fuels 64, 65
    biofuels 65
    crude oil 64
    greenhouse gases 65
    hydrogen fuel cells 64
Newton, Isaac 117
Noddack, William 77
nuclear fusion 113

Oersted, Hans Christian 130
Archicebus 80, 81
On the Origin of Species 33
oxides of nitrogen 62, 63

pacemakers 107
palladium 67
particles 144, 145
    accelerators 144
    Higgs boson 145
    Large Hadron Collider 144, 145
    quarks 145
    semiconductors 145
particulates (diesel) 63
pasteurisation 17
pathogens 24
pathology 48, 49
    bite marks 49
    pathologists 48
    teeth 49
peer review 33, 80, 131
penicillin 26–28
Penicillium notatum 27
Periodic Table 76, 77
    catalysts 77
    elements 76
    rhenium 77
petrol (fossil fuel) 64
Phipps, James 23
phones 102, 103
    electromagnetic spectrum 103
    microwaves 103
    radio waves 103
    waves 103
planets (detection) 136, 137
plants (cloning) 13
plasma 44, 45
Plato 117

platelets 44
platinum 67
plum-pudding model 72, 73
polar orbit 123
pollen 39
polonium 126
polydactyly 7
poly(propene) 94, 95
position (detection) 140, 141
premature babies 106
projectile motion 110
protons 144
Proxima Centauri 139, 141
Ptolemy 118
pulse rate 106
Punnett squares 5

quarks 145

radar (RAdio Detection And Ranging) 140
radio waves
    AM and FM radio 142
    diffraction 143
    phones 103
radioactivity 126–129
    alpha rays 127
    beta rays 127
    cancer 128, 129
    gamma camera 129
    gamma rays 127
    half-life 127
    mutations 128
    radiation 126, 127
    radiotherapy 129
    radium 126
    sterilisation 129
radiotherapy 129
radium 126, 128
reaction time 110, 111
reactivity series 93
red blood cells 44, 45
rennet 16
resistance (antibiotics) 28
resolution (microscopy) 38
retrograde motion 118, 119
rhenium 77
rhodium 67
rigor mortis 46
rooting powder 13
rust 93

Rutherford, Ernest 73

Saccharomyces cerevisiae 14
safety
    Formula 1 111
    nanoparticles 60, 61
    sunscreen 60, 61
scandium 77
scarlet fever 27
scientific journals 80
Search for Extra Terrestrial Intelligence (SETI) 138
sedimentary rocks 79
seed banks 35
selective breeding 8, 9
semiconductors 145
SETI (Search for Extra Terrestrial Intelligence) 138
shock-absorbing materials 125
side-effects 25
sisal 94
smallpox 23
Smith, William 79
solar cells 113
solar flares 124
Solar System 121, 122, 139
spacecraft and satellites 122, 123
    geostationary orbit 123
    low Earth orbit 123
    polar orbit 123
sports 110, 111
    projectile motion 110
    reaction time 110, 111
    safety 111
Sputnik 1 122
stain removal 18
stationary phase (chromatography) 89
sterilisation 29
strata (fossils) 79
Streptococcus thermophiles 17
submarines 132
sugar fermentation 15
Sun 119
    heliocentric model 119
superbugs 28, 29

Tacke, Ida 77

teeth (pathology) 49
telescopes 136
temperature
    body 47
    enzymes 19
theory of evolution 32, 33
thermistors 106
Thomson, J. J. 72
thymine (T) 30
time of death 46, 47
total internal reflection 108
transmitters 132

ultrasound 109
universal indicator 86
Universe
    aliens 100
    origin 100, 101
urine 48

vaccines 22–25
    antibodies 24
    cowpox 23
    smallpox 23
ventilators 107

Wallace, Alfred 31
waste substances 66
water filters 125
Watson, James 31
waves 133
    electromagnetism 133
white blood cells 44, 45
whorl (fingerprinting) 41
Wilkins, Maurice 31
wind turbines 112
wine 15
wool 39, 94
woolly mammoths 34

X-rays 108
Xijun, Ni 80

yeast 14, 15
yoghurt 16, 17

Zamith, Sebastien 54
zinc oxide 60

The Periodic Table

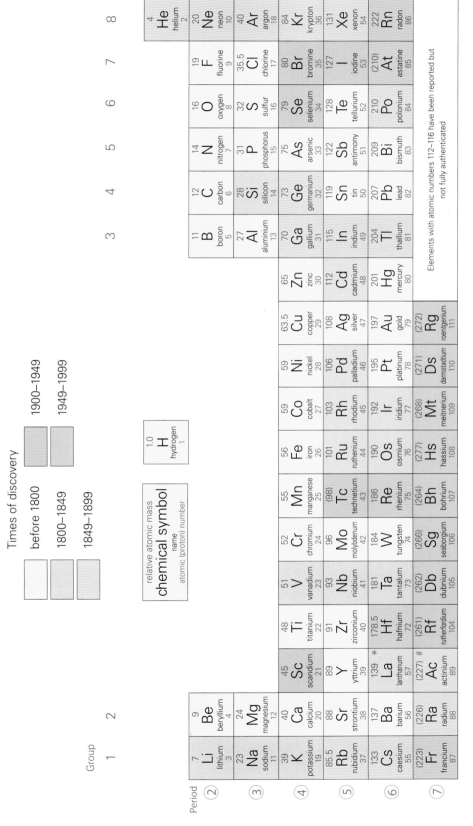

Great Clarendon Street, Oxford, OX2 6DP, United Kingdom

Oxford University Press is a department of the University of Oxford. It furthers the University's objective of excellence in research, scholarship, and education by publishing worldwide. Oxford is a registered trade mark of Oxford University Press in the UK and in certain other countries

© Oxford University Press 2014

The moral rights of the authors have been asserted

First published in 2014

All rights reserved. No part of this publication may be reproduced, stored in a retrieval system, or transmitted, in any form or by any means, without the prior permission in writing of Oxford University Press, or as expressly permitted by law, by licence or under terms agreed with the appropriate reprographics rights organization. Enquiries concerning reproduction outside the scope of the above should be sent to the Rights Department, Oxford University Press, at the address above.

You must not circulate this work in any other form and you must impose this same condition on any acquirer

British Library Cataloguing in Publication Data
Data available

978-0-19-839258-3

10 9 8

Paper used in the production of this book is a natural, recyclable product made from wood grown in sustainable forests. The manufacturing process conforms to the environmental regulations of the country of origin.

Printed in China by Golden Cup

Acknowledgements

The publisher and the authors would like to thank the following for permissions to use their photographs:

Cover image: Shaskin/Shutterstock; **p2**: Rido/Shutterstock; p2: (background) Power And Syred/Science Photo Library, (tl) Mafaldita /iStockphoto; p3: (tl1) Elena Elisseeva/Shutterstock, (tl2) RichLegg/iStockphoto, (tc1) JoeGough/Bigstock, (tc2) CristinaMuraca / Shutterstock, (tr1) Dr. Gopal Murti/Science Photo Library, (tr2) Mafaldita /iStockphoto, (b) Power And Syred/Science Photo Library; p4: Mafaldita /iStockphoto; p6: skynesher/iStockphoto; p7: Science Photo Library; p9: (l) 101cats/iStockphoto, (r) AndrewJohnson/iStockphoto; p10: Getty Images/Getty Images News /Getty Images); p11: (tl) HelleM/Shutterstock, (tr) tomch/iStockphoto, (b) luismmolina /iStockphoto; p12: G.CIGOLINI/De Agostini Picture Library/Getty Images; p13: (tr) Geoff Kidd/Science Photo Library, (bl) Philippe Plailly/Science Photo Library; p14: (t) Power And Syred/Science Photo Library, (b) Martyn F. Chillmaid/Science Photo Library; p15: (t) mattjeacock / iStockphoto, (b) Skyhobo/iStockphoto; p16: (t) JoeGough/Bigstock, (m) seraficus / iStockphoto, (b) RIA NOVOSTI/Science Photo Library; p17: Scimat/ Science Photo Library; p18: (t) CGissemann / Shutterstock, (b) Oliver Hoffmann/Shutterstock; p20: (background) Power And Syred/Science Photo Library, (tr)Getty Images / Getty Images News / Getty Images, (mr) JoeGough/Bigstock; p22: (t) Elena Elisseeva/Shutterstock, (b) muchemistry /iStockphoto; p23: (tr) Georgios Kollidas/Shutterstock, (bl) Nypl/Science Source/Science Photo Library; p24: Eric Grave/Science Photo Library; p26: (tl) Dr p. Marazzi/Science Photo Library, (bl) James King-Holmes/Science Photo Library, (br) St Mary's Hospital Medical School/Science Photo Library; p27: (tr) kmitu/Bigstock, (bl) ksass / iStockphoto; p28: Dr Kari Lounatmaa/Science Photo Library; p29: (tl) Life In View/Science Photo Library, (bl) Dougberry/iStockphoto; p31: Science Photo Library; p32: (l) National Library Of Medicine/Science Photo Library, (r) Science Photo Library; p33: Science Photo Library; p34: (t) Natural History Museum, London/Science Photo Library, (b) gutang /iStockphoto; p35: (tr) keiichihiki / iStockphoto, (bl) Frans Lanting, Mint Images / Science Photo Library; p36: (t) gutang / iStockphoto, (m) Dr Kari Lounatmaa/Science Photo Library; p38: RichLegg/iStockphoto; p39: (tl) alanphillips/iStockphoto, (tc) Dr.Jeremy Burgess/Science Photo Library, (tr) AMI Images/Science Photo Library, (mr) Science Photo Library, (br) Power And Syred/Science Photo Library; p40: Dorling Kindersley/Getty Images; p41: (t) Joe Belanger/ Shutterstock, (m) ABDesign /iStockphoto; p42: (t) BackyardProduction/iStockphoto, (b) red_moon_rise/iStockphoto; p43: zmeel /iStockphoto; p44: (l) jimbycat /iStockphoto, (b) Steve Gschmeissner/Science Photo Library; p45: RMAX/iStockphoto; p46: barclayboy/iStockphoto; p47: (t) rlindo71 /iStockphoto, (m1) dabjola/Shutterstock, (m2) Eric Isselee/Shutterstock, (b) Henrik Larsson/Shutterstock; p48: (l) mariusFM77/iStockphoto, (br) zmeel /iStockphoto; p49: (t) zokara/iStockphoto, (bl) Benne Ochs/fstop/Corbis; p50: (background) AMI Images/Science Photo Library, (tr) Science Photo Library, (mr) zokara/iStockphoto, (b) paulthepunk / iStockphoto; p52: (background) belterz/iStockphoto, (t) Science Photo Library, (b) Clive Streeter / Dorling Kindersley/Getty Images; p53: (tl1) Martin Brayley/Dreamstime, (tl2) FauxCaster /iStockphoto, (tc1) alexandrumagurean/iStockphoto, (tc2) davidf/iStockphoto, (tr1) Sailorr / Shutterstock, (tr2) Jiri Slama/Shutterstock, (b) EdeWolf/iStockphoto; p54: (t) Patrick Dumas/Look At Sciences/Science Photo Library, (b) SunChan/iStockphoto; p55: (t) Royal Institution Of Great Britain / Science Photo Library, (b) David Mccarthy/Science Photo Library; p56: (tl) BanksPhotos/iStockphoto, (mc) Tyler Boyes /Shutterstock, (mr) Eye Of Science/Science Photo Library, (bl) labsas/iStockphoto; p58: Dr p. Marazzi/Science Photo Library; p60: (t) MarkSwallow/iStockphoto, (b) Picsfive / Shutterstock; p61: (t) ssuaphoto/iStockphoto, (b) dra_schwartz/iStockphoto; p63: SchmitzOlaf /iStockphoto; p64: (t) FauxCaster/iStockphoto, (bm) Lex Bartel/Science Photo Library, (br) Robert Timoney / Alamy; p65: (tl) santosha/iStockphoto, (tm) PeterAustin/iStockphoto, (tr) Martin Bond/Science Photo Library; p66: Astrid & Hanns-Frieder Michler/Science Photo Library; p67: (l) Clive Streeter/Dorling Kindersley/Getty Images, (r) StockThings/Shutterstock; p68: davidf / iStockphoto; p69: FocusTechnology / Alamy; p70: (background) FauxCaster/iStockphoto, (tr) Eye Of Science/Science Photo Library, (mr) labsas/iStockphoto; p72: (t) pixhook/iStockphoto, (b) Science Photo Library; p74: Michal Zak / Shutterstock; p75: Dorling Kindersley/Getty Images; p76: Ria Novosti/Science Photo Library; p77: (tr) Science Photo Library, (bl) Science Source/Science Photo Library; p78: (l) IgorGolovniov/Shutterstock, (m) John Reader/Science Photo Library; p80: (t) Dr. Xijun NI, (b) Dr. Xijun NI; p81: holgs/iStockphoto; p82: (background) John Reader/Science Photo Library, (tl) holgs/iStockphoto, (ml) Dorling Kindersley/Getty Images; p84: (t) Plougmann / iStockphoto, (b) upsidedowndog/iStockphoto; p86: (l) Martin Brayley/Dreamstime, (r) Jiri Slama/Shutterstock; p87: (tr) Wdeon/iStockphoto, (m) ohan Larson/Shutterstock; p88: (l) zmeel / iStockphoto, (r) Sailorr/Shutterstock; p90: Volker Steger/Science Photo Library; p92: (tl) davidf / iStockphoto, (bl) Mauro Carli/Shutterstock, (br) cobalt/iStockphoto; p94: (tl) PelageyaKlubnikina / iStockphoto, (r) pierivb/iStockphoto, (ml) alexandrumagurean/iStockphoto, (bl) studiovd1 / iStockphoto; p95: Photogalia/iStockphoto; p96: (r) belterz/iStockphoto; p97: (tl) Masterfile, (tr) Rich Legg/iStockphoto, (bl) csreed/iStockphoto; p98: (background) belterz/iStockphoto, (tr) alexandrumagurean/iStockphoto, (mr) Jiri Slama/Shutterstock; p100: (background) Nasa/Esa/Stsci/Science Photo Library, (t) Arno Massee/Science Photo Library, (m) Soho/ESA/NASA/Science Photo Library; p101: (tl1) NASA/Science Photo Library, (tl2) Pyshnyy Maxim Vjacheslavovich /Shutterstock, (tc1) Fuse/Getty Images, (tc2) NASA/Science Photo Library, (tr1) NASA/Science Photo Library, (tr2) Ted Kinsman/Science Photo Library, (b) Klaus Guldbrandsen/Science Picture Library; p102: (tl) Jeremy Horner/Corbis, (tr) Sam Ogden/Science Photo Library; p103: Claus Lunau/Science Photo Library; p104: (t) Jiri Hera/Shutterstock, (b) janda75/iStockphoto; p105: (t) Edward Kinsman/Science Photo Library; p106: (l) AJ photo/hop. Americain/Science Photo Library; p107: (t) Dr P. Marazzi/Science Photo Library, (ml) Michael Donne/Science Photo Library, (mr) Deep Light Productions/Science Photo Library, (b) LISSAC/Science Photo Library; p109: Arno Massee/Science Photo Library; p110: (tl) lev radin/Shutterstock, (tr) Kenneth Eward/Science Photo Library, (bl) Howard Kingsnorth/Science Photo Library; p111: Pedro Ugarte/AFP/Getty images; p112: Jesus Keller/Shutterstock; p113: European Space Agency/J. Huart/Science Photo Library; p114: (background) janda75/iStockphoto, (tr) Edward Kinsman/Science Photo Library, (mr) Jesus Keller/Shutterstock; p116: (l) Yuttasak Jannarong / Shutterstock, (c) Royal Astronomical Society/Science Photo Library /Science Photo Library, (r) Sheila Terry/Science Photo Library; p117: (r) NASA/Science Photo Library; p118: (b) Sheila Terry/Science Photo Library; p120: (l) Victor De Schwanberg/Science Photo Library, (r) Stocktrek/Photodisc/Getty Images; p121: (l) Mark Garlick/Science Photo Library, (r) MaraZe/Shutterstock; p122: (b) Julian Baum/Science Photo

Library; p124: (t) NASA/Science Photo Library, (b) NASA/Science Photo Library; p125: (tr) Soho/ESA/NASA/Science Photo Library, (tl) Pyshnyy Maxim Vjacheslavovich /Shutterstock, (tc) Photobac/Shutterstock, (ml) YanLev/Shutterstock, (bl) FamVeld /Shutterstock, (bc) Tomislav Pinter/Shutterstock; p126: (tl) Martyn F. Chillmaid/Science Photo Library, (bl) W. F. Meggers Collection/American Institute of Physics/Science Photo Library, (r) Science Photo Library; p128: (tl) Public Health England/Science Photo Library, (bl) Public Health England/Science Photo Library, (br) Kaspri/Shutterstock; p129: (tr) Lack-O'Keen/Shutterstock; p130: (t) Giphotostock/Science Photo Library, (b) British Library/Science Photo Library; p131: (tr) Awe Inspiring Images/Shutterstock, (bl) Sheila Terry/Science Photo Library; p132: (tl) Iurii/Shutterstock, (bl) Christopher Wood/Shutterstock, (br) NYPL/Science Source/Science Photo Library; p133: Fuse/Getty Images; p134: (background) Mark Garlick/Science Photo Library, (tr)Public Health England/Science Photo Library, (mr) Martyn F. Chillmaid/Science Photo Library; p136: Amy Walters/Shutterstock; p137: (tr) NASA/Ames/JPL-Caltech/Wikipedia, (mr) Nasa/Esa/Stsci/Science Photo Library; p138: (t) David Parker/Science Photo Library, (b) NASA/Science Photo Library; p139: Science Photo Library; p140: (t) Ted Kinsman/Science Photo Library, (b) Ed Young/Science Photo Library; p142: DmitriMaruta/Shutterstock; p144: (t) Galina Barskaya/Shutterstock, (b) Adam Hart-Davis/Science Photo Library; p145: (t) CERN/Science Photo Library, (b) Thomas Mccauley, Lucas Taylor/Cern/Science Photo Library; p146: (background) Thomas Mccauley, Lucas Taylor/Cern/Science Photo Library, (tr) NASA/Science Photo Library, (mr) Amy Walters/Shutterstock

Artwork by Phoenix Photosetting and Q2A Media

Although we have made every effort to trace and contact all copyright holders before publication this has not been possible in all cases. If notified, the publisher will rectify any errors or omissions at the earliest opportunity.

Links to third party websites are provided by Oxford in good faith and for information only. Oxford disclaims any responsibility for the materials contained in any third party website referenced in this work.